Nature Nearby

THE WILEY NATURE EDITIONS

AT THE WATER'S EDGE
Nature Study in Lakes, Streams, and Ponds,
by Alan M. Cvancara

WALKING THE WETLANDS
A Hiker's Guide to Common Plants and Animals of Marshes, Bogs, and Swamps,
by Janet Lyons and Sandra Jordan

MOUNTAINS
A Natural History and Hiking Guide,
by Margaret Fuller

THE OCEANS
A Book of Questions and Answers,
by Don Groves

WILD PLANTS OF AMERICA
A Select Guide for the Naturalist and Traveler,
by Richard M. Smith

FORESTS
A Naturalist's Guide to Trees and Forest Ecology,
by Laurence C. Walker

NATURE NEARBY

An Outdoor Guide to 20 of America's Cities

Bill McMillon

Wiley Nature Editions

John Wiley & Sons

New York Chichester Brisbane Toronto Singapore

Library of Congress Cataloging-in-Publication Data

McMillon, Bill, 1942-
Nature nearby : an outdoor guide to 20 of America's cities / Bill McMillon.
p. cm. — (Wiley nature editions)
Includes bibliographical references.
ISBN 0-471-62339-3 (pbk.)
1. Natural areas—United States—Guide-books. 2. Cities and towns—United States—Guide-books. I. Title. II. Series.
QH76.M3847 1990
508.73'09173'2—dc20 89-70760
CIP

Printed in the United States of America

90 91 10 9 8 7 6 5 4 3 2 1

"Delight is a weak term to express
the feelings of a naturalist
who, for the first time,
has wandered in a Brazilian forest."

CHARLES DARWIN
Voyage of the Beagle

To David,

who suggested the book

and made sure I finished,

even if not on time.

Contents

SOUTHWEST
151

FAR WEST
193

Preface

Darwin was undoubtedly right about the enjoyment of wandering in a Brazilian forest for the first time, but few amateur naturalists have that opportunity. More likely their wanderings are among the shadows of towering buildings in the asphalt jungles of America's cities, or in some park or preserve nearby.

Being the odd sort they are, and I must count myself among their numbers, amateur naturalists quickly tire of the glitzy, human-made attractions that abound in cities. Instead, they yearn for small pieces of nature among the concrete and steel to help satisfy an inner craving for contact with the real rather than the artificial.

While some may not understand what these naturalists are searching for, others will. They agree by paraphrasing Darwin, saying, "Delight is a weak term to express the feelings of an amateur naturalist who, for the first time, finds an outpost of nature amidst the concrete and asphalt of a city."

Such outposts are hard to come by, especially in cities far from the naturalist's home. Convention and visitors bureaus in most large cities are generally ill informed about nature organizations and sites. When an outing or organization is discovered and enjoyed, it most likely comes by serendipity. An unpublicized arboretum or an overlooked section of tide flats is usually stumbled across during other activities. But, many metropolitan areas do have nature centers, natural areas, and nature preserves nearby. Visiting naturalists can find escape, for an hour or a day, from the sterile conrete-and-steel monoliths and car-laden freeways that fill the cities.

It is difficult to find these sites and organizations because little has been written about them. This guidebook was developed to help fill that void. It is now easier for traveling naturalists to find nature-related activities when they are far from home, and want to enjoy a brief respite from the demands of the city.

For the book, the country is divided into five rather arbitrary regions, the Northeast, Southeast, Midwest, Southwest, and Far West. Four representative cities are included in each section. Activities in or near the cities are emphasized, but interesting and unusual activities up to two hours away from the city are also included. For some regions this means that activities from other nearby cities may be included. For cities such as New York and Philadelphia, there is some overlap because of their proximity.

While most naturalists would prefer to spend their time exploring the outdoors, there are always circumstances that preclude that. Therefore indoor activities that are of interest to naturalists are included for each city.

Nature was not designed with street numbers, ZIP codes, and political boundaries. As a result, it is often difficult to obtain specific addresses for individual sites. To help readers locate sites, general directions are given. To help you contact agencies responsible for administering those sites, telephone numbers and specific addresses are given when available. These are often mailing addresses for agency headquarters far from the actual site. Specific directions may be obtained from the headquarters.

As with any directory or guidebook where information is collected over a period of time from a wide number of sources, there may be some omissions and inaccuracies. I hope, however, these have been kept to a minimum, and that all readers will find the information a valuable resource that will make their visits to these cities more enjoyable.

Introduction

Nature Nearby is arranged so even novice naturalists can find outings and activities that are satisfying, and give pleasant and informative breaks from the usual tourist or business fare. It covers 20 of America's most visited urban areas, gives details of many activities that are available in each region, notes organizations that sponsor naturalist activities such as hikes and lectures, and lists natural history museums, nature centers, arboretums, botanical gardens, and natural areas that are open to the public.

Since there is not enough space to thoroughly discuss all the activities available in each city, most points of interest are listed with simple descriptions of location and available activities. Outstanding or unusual outings for each city are highlighted, however, to give you a better feel of what can be found in the area.

In addition to a directory of organizations and nature activities available in each city, a short bibliography of resource materials is included so that you can read about the natural history of the region before you visit. This will enhance your enjoyment of any activities you participate in while there.

A short overview of the climate, geography, and general natural history of each city is given so first-time visitors will know what to expect when they reach their destination, and can prepare for contingencies such as foul weather. Although many of you are experienced naturalists and travelers, others are novices and may find the following hints helpful.

Planning Ahead

This guidebook will be most helpful if you have the opportunity to plan your trip ahead of time. It can also be used when you must go to a new city with little advance notice.

For the former, plans can be made and advance registration for outings can be arranged before you leave. This will give you an opportunity to include any special articles such as camera, hiking boots, clothing, and bird books, if they are appropriate for your planned outings.

For the latter, the book is an excellent resource that you can carry with you to help locate activities that can be enjoyed on the spur of the moment when you find time on your hands. Simply look up your favorite activities in the city, and find those that can be enjoyed without advance registration or equipment.

Group Activities

Many of the organizations listed offer guided outings that are often popular and require advance registration. They are crowded for the very reason you will want to join them; they give participants opportunities to closely observe some part of nature near a major urban area.

All the organizations listed throughout the book sponsor at least an occasional outing or lecture. For a listing of current activities contact those organizations that specialize in your areas of interest. Do this far enough ahead of time so you can get confirmation of your registration, and prepare for any special requirements and equipment.

Keep in mind that many outings are dependent upon the weather, and that they are often cancelled or rescheduled. If you are visiting during a season with unpredictable weather, have alternate plans in case your outing is cancelled.

Getting Around

Most of the metropolitan areas in this guide have at least adequate public transportation, so you can visit many of the sites listed without worrying about car rentals. Some, such as Los Angeles, are so large and scattered and have relatively scarce transit systems. In such instances you will either have to rent a car to reach the activities, or check with the sponsoring organizations to see if they have any car pool arrangements for their outings. Some outings led by local organizations may require use of private transportation.

Outings With Children

Most of the activities listed in this guide are appropriate for children. Some are even designed especially for them, such as junior museums. Some organizations, however, do not allow children below a certain age on their outings. Always check on this before sending in registration materials, particularly if you have preteens.

Special Equipment

The idea of special equipment for nature outings in and around cities may seem outrageous, but there are some items that add to the enjoyment of various nature activities. Those of you interested in bird watching will find binoculars and bird guides indispensable. Cameras and special lenses are always useful for those who wish to record segments of their adventures. Comfortable hiking clothes and shoes appropriate to the local weather, and this often means rain gear, will make all outings more enjoyable.

Natural and Unnatural Hazards

With the earthquake of San Francisco and Hurricane Hugo along the East Coast in 1989, nature once again demonstrated that it was not to be controlled. Anyone who spends a lot of time exploring nature is well aware that there are many natural hazards that can lead to disaster if not taken into account when outings are planned.

No region of the country is without natural hazards. The South, from the Eastern Seaboard to Texas, is prone to hurricanes and tornadoes. The East has an occasional hurricane, as well as blizzards. The Midwest has tornadoes and blizzards. The Southwest has the deadly heat of the desert and flash floods. And the Far West has earthquakes and volcanoes. All regions have hazards such as swamps where intruders can easily get lost, mountains where unexpected cold snaps can lead to snow, or deserts where visitors must be careful about losing their way in areas without water or shade.

Naturalists are generally well aware of these hazards, but must sometimes be reminded of dangers in regions that are unfamiliar to them. What many are not as aware of are the unnatural hazards that may be encountered as explorations are made in and around major metropolitan areas.

The most common of these hazards comes from other humans; ones who have little respect for their fellow beings and are likely to take advantage of a chance encounter in an isolated park or trail. To avoid such dangers you should always inquire about the dangers in a new park or nature reserve, and ask if it is wise to venture out alone. With adequate precautions, both natural and unnatural hazards can be minimized so the many nature activities in this guide can be enjoyed without fear.

Nature Is Where You Find It

Nature can be found in strange places in cities. University campuses often have arboretums where native birds and local vegetation are encouraged, and cemeteries can be ideal bird-watching sites. Vacant lots are a good site for wildflower displays in the spring, as well as freeways, railway, and rapid transit right-of-ways. Dumps and sewage treatment plants are frequently feeding sites for a variety of birds.

You have to be resourceful to enjoy nature in urban areas. You should also have some knowledge of what might be available in the city you are visiting. This guide book is just an introduction. You will find others as you visit the cities, and make plans for future visits. The following reading list and the ones in the later sections will help you expand your knowledge about the natural world, and how it can be found anywhere, including the nation's major cities.

Further Reading

Information about a particular region or subject always adds to the enjoyment of any nature outing. The following list of general books covers large regions of the United States. You will find these books informative, whether you are sitting in your own backyard, visiting nature sites in or near large cities, or hiking deep into a wilderness area. Specific reading lists are included for each region and city to help you enjoy your nature outings there.

Angel, H., and Pat Wolseley. *The Water Naturalist.* New York: Facts on File, 1982.

Audubon Society Field Guides. New York: Alfred A. Knopf. This series includes guides to a wide variety of flora and fauna in the United States. All the regions included in *Nature Nearby* are covered.

Brainerd, John W. *The Nature Observer's Handbook: Learning to Appreciate Nature As You Travel.* Chester, CT: Globe Pequot, 1986.

Brown, Lauren. *Grasses: An Identification Guide.* Boston: Houghton Mifflin Co., 1979.

Clark, William S., and Brian K. Wheeler. *Hawks.* Boston: Houghton Mifflin Co., 1987.

Covell, Charles V., Jr. *A Field Guide to the Moths of Eastern North America.* Boston: Houghton Mifflin Co., 1984.

Crockett, Lawrence J. *Wildly Successful Plants: A Handbook of North American Weeds.* New York: Collier, 1977.

Dunne, Pete, David Sibley, and Clay Sutton. *Hawks in Flight.* Boston: Houghton Mifflin Co., 1988.

Ehrlich, Paul R., David S. Dobkin, and Darryl Wheye. *The Birder's Handbook: A Field Guide to the Natural History of North American Birds.* New York: Fireside, 1988.

Elias, Thomas S. *The Complete Trees of North America.* New York: Van Nostrand Reinhold Co., 1980.

Farrand, John, Jr. *An Audubon Handbook: How to Identify Birds.* New York: McGraw-Hill Book Co., 1988.

———. *An Audubon Handbook: Western Birds.* New York: McGraw-Hill Book Co., 1988.

Garber, Steven D. *The Urban Naturalist.* New York: John Wiley & Sons, 1987.

Geffen, Alice M. *A Birdwatcher's Guide to the Eastern United States.* Woodbury, NY: Barron's, 1978.

George, Carl J., and Daniel McKinley. *Urban Ecology in Search of an Asphalt Rose.* New York: McGraw-Hill Book Co., 1974.

Harrison, Peter. *Seabirds: An Identification Guide.* Boston: Houghton Mifflin Co., 1983.

Hayman, Peter, John Marchant, and Tony Prater. *Shorebirds: An Identification Guide to the Waders of the World.* Boston: Houghton Mifflin Co., 1986.

Heckscher, August. *Open Spaces: The Life of American Cities.* New York: Harper & Row, 1977.

Kinkead, Eugene. *Wildness Is All Around Us, Notes of an Urban Naturalist.* New York: Dutton, 1978.

Knobel, Edward. *Field Guide to the Grasses, Sedges, and Rushes of the United States.* New York: Dover, 1980.

Korling, Torkel, and Robert O. Petty. *Wild Plants in Flower: Eastern Deciduous Forest.* Chicago: Chicago Review Press, 1978.

Madge, Steve, and Hilary Burn. *Waterfowl: An Introduction to the Ducks, Geese and Swans of the World.* Boston: Houghton Mifflin Co., 1988.

McPhee, John. *In Suspect Terrain.* New York: Farrar, Straus, & Giroux, 1983.

Mitchell, John Hanson. *A Field Guide to Your Own Back Yard.* New York: Norton, 1985.

Neiring, William. *The Life of a Marsh.* New York: McGraw-Hill Book Co., 1966.

Newcomb, Lawrence. *Newcomb's Wildflower Guide.* Boston: Little, Brown & Co., 1977.

Orr, Robert T. *Animals in Migration.* New York: Macmillan Co., 1970.

Perry, John and Jane G. *The Random House Guide to Natural Areas of the Eastern United States.* New York: Random House, 1980.

Peterson Field Guide Series. Boston: Houghton Mifflin Co. This is another series that has in-depth guides on a wide variety of nature subjects.

Pettingill, Olin Sewall, Jr. *A Guide to Bird Finding East of the Mississippi.* 2nd ed. New York: Oxford University Press, 1977.

Robbins, Chandler S., Bertel Broun, and Herbert Zim. *Birds of North America.* New York: Golden Press, 1966.

Russell, Helen Ross. *City Critters.* Cortland, NY: American Nature Society, 1975.

Scott, Shirley, ed. *Field Guide to the Birds of North America.* 2nd ed. Washington, DC: National Geographic Society, 1983.

Stebbins, Robert C. *A Field Guide to Western Reptiles and Amphibians.* Berkeley, CA: University of California Press, 1966.

Teal, John and Mildred. *The Life and Death of the Salt Marsh.* New York: Ballantine Books, 1969.

Thomas, Bill. *Eastern Trips and Trails.* Harrisburg, PA: Stackpole Books, 1975.

Tilden, James W., and Arthur Clayton Smith. *A Field Guide to Western Butterflies.* Boston: Houghton Mifflin Co., 1986.

Vankat, John L. *Natural Vegetation of North America.* New York: John Wiley & Sons, 1979.

Watts, M.T. *Master Tree Finder.* Berkeley, CA: Nature Study Guide, 1963.

White, Richard E. *A Field Guide to the Beetles of North America.* Boston: Houghton Mifflin Co., 1983.

NORTHEAST

Boston
New York
Philadelphia
Washington, DC

The early European settlers of the United States came to the Northeast. There they tamed the wilderness and built the country's first cities. As these cities grew into large metropolitan areas, the region became almost devoid of natural areas undisturbed by human settlers.

Enough open space was left, however, for some natural habitats to survive. Today, visitors to Boston, New York, Philadelphia, and Washington, DC, can find a variety of naturalist activities in which to participate. As with other sections of the country, some of these cities have more to offer than others.

New York has an active outdoor community with wildlife refuges within the city limits and several organizations that lead outings. Boston, on the other hand, has comparatively fewer organizations. Little is left of the vast natural world that existed in and around Boston Harbor before the Pilgrims landed at Plymouth Rock.

Natural History of the Northeast

Water, in the form of glaciers, waves, and rivers, has helped shape the geography of the eastern section of the United States. Major inlets and bays along the coast formed, and early settlers built their cities there. Boston and New York grew up around large natural harbors; Philadelphia sits between two major rivers; and Washington was built on the Potomac River, which empties into the Chesapeake Bay, one of the great natural areas along the east coast.

The proximity of these cities to large bodies of water tempers their climates somewhat, and makes them more moderate than those farther inland. Visitors to the region must still contend with hot, muggy summers and cold winters. Spring and fall are the best times to enjoy naturalist's activities, although there is plenty to do during the other two seasons.

Historically, native habitats of the region included extensive growths of hardwood and riparian forests, swamps, marshes, and tide flats, as well as large dunes and wild seashores. These were home to numerous native birds and mammals.

The forests, marshes, and tide flats have largely disappeared around the cities. Boston has only a small remnant of the salt marsh that once surrounded Boston Harbor. Much of the original wildlife of the region has met a similar fate.

Because uncontrolled expansion of cities in this region has so drastically reduced the possibilities of experiencing nature firsthand, the natural sites that can be found in or near the urban areas become more important to residents and visitors.

To best enjoy the natural sites near northeastern cities, you should read about salt marsh, seashore, hardwood forest, and riparian life zones. These are the most common natural sites left.

General books on these subjects are found in the reading list from the Introduction. The following list of books includes specific information on the Northeast.

Further Reading

Babcock, Harold L. *Turtles of the Northeastern United States.* New York: Dover, 1971.

Berman, Steve. *The Northeastern Outdoors.* Boston: Stone Wall Press, 1978.

Garvey, Edward B. *Hiking Trails in the Mid-Atlantic States.* Matteson, IL: Greatlakes Living Press, 1976.

Henley, Thomas A., and Neesa Sweet. *Hiking Trails in the Northeast.* Matteson, IL: Greatlakes Living Press, 1976.

Klimas, John E., and James A. Cunningham. *Wildflowers of Eastern America.* New York: Galahad Books, 1981.

Raymo, Chet and Maureen. *A Geological History of the Northeastern United States.* Chester, CT: Globe Pequot, 1989.

Prentice, Thurlow Merrill. *Weeds and Wildflowers of Eastern North America.* Salem, MA: Peabody Museum, 1973.

Boston

Human history, not natural history, is the primary tourist activity in Boston. There is plenty of human history for visitors to enjoy in walks around the city. Boston's early residents played an integral part in the formation of the 13 colonies and the United States. Consequently the major efforts of Boston's National Park Service center around the Freedom Trail, which emphasizes the activities of early Boston residents.

Natural history activities are much harder to come by in the region. Massachusetts is a small state with no large natural areas, especially in the southeastern part where Boston is located. The state has no national parks or forests, and less than 15,000 acres of wildlife refuges. It does manage several hundred parks and forests, but not all of these can be considered natural areas. Even fewer of those near Boston qualify.

ABOVE: St. Georges Island, one of seven islands in the Boston Harbor Islands State Park, offers hikes over isolated hills, as well as tours of an old fort. Photo courtesy of Metropolitan District Commission.

In Boston, little effort was made to preserve natural areas as the city grew. The hills that once jutted above the harbor have mostly been leveled and used as landfill, which in turn, has displaced most of the salt marshes that once rimmed the bay.

The Charles River, Massachusetts Bay, Cape Cod Bay, and Cape Cod are the natural features that dominate Boston's geography. The Appalachian Mountains that traverse western Massachusetts, known as the Berkshire Hills, are about three hours from Boston.

The forests near Boston are generally secondary forests, primarily deciduous with some conifers, that offer a mixture of shrubs, meadows, and forest. These are favorable habitats for many small mammals and birds.

Closer to the Atlantic Ocean, particularly on Cape Cod and the Boston Harbor Islands, there is a very different habitat. There the seashore is home to a wide variety of shorebirds and waterfowl, and the dunes are covered with low shrubs and pines.

Cape Cod is the prominent feature of the Massachusetts coastline, and its land mass deflects the Gulf Stream on its northward journey. The sharpest division of ocean life forms that has ever been found occur here, as life south of the cape is completely different from that to the north. The seashores, both near Boston and on Cape Cod, are overly crowded during the summer months. Attempts to find an enjoyable nature outing on or near them during hot periods is a lesson in frustration. Fall and winter days are frequently sunny and mild in the region, as the weather is moderated by the Gulf Stream, and the seashore is an excellent choice for a nature outing during those times.

Climate and Weather Information

In comparison to the inland regions of New England, Boston has a moderate climate. Winter temperatures seldom drop much below freezing, snow fall is moderate, and daytime temperatures average in the high-30s and low-40s even in mid-winter.

Highs in the spring and fall are likely to be in the 60s, and offer the most pleasant days for exploring nature. Summers are hotter—around 80—and are humid. This is the weather that brings the hordes to the seashore, and floods the riverside parks in the city with crowds of sun worshipers. Boston's annual rainfall is high, and most falls during the spring and summer months.

Getting Around Boston

Driving in Boston is reputed to be some of the worst in the nation. The streets do not follow a conventional grid pattern, and it is easy to lose yourself. Parking is limited. Rent a car only if it is necessary. Some drivers prefer to stay on the outskirts of the city and rid themselves of rental cars when they return.

On the plus side, you can use the excellent public transit system to get around in the city, and to get to the edge where you can rent a car. The fares on the public transit system are inexpensive. The system includes subways, trolleys, and bus lines. Taxis are easily found, and fares compare with other major cities. One word of caution: Most of the nature activities listed for Boston are away from the city center. Cape Cod, for example, is about two hours away. You will probably have to rent a car to reach most of the activities and sites mentioned.

Indoor Activities

The climate in the Northeast is unpredictable enough to keep even the most ardent naturalist inside sometime during the year. The nature centers, organizations, and museums have nature-related activities and exhibits that are indoors and that can be enjoyed during inclement weather.

Boston University Observatory
725 Commonwealth Ave., Boston, MA 02215—617-353-2630

This university observatory has an open telescope year-round, 9:00 P.M.–10:00 P.M. in summer and 8:00 P.M.–9:00 P.M. in winter. Observatory staff members are available to assist you. The Observatory is included in this section, although the need for clear skies precludes this activity during inclement weather.

Cape Cod Museum of Natural History
Drawer R State Route 6A, Brewster, MA 02631—508-896-3867

This museum on Cape Cod presents native flora and fauna. Additionally, there are outdoor nature trails. It is open Monday–Saturday from 9:00 A.M. to 4:30 P.M. and Sundays from 12:30 P.M. to 4:30 P.M. during the summer; Tuesday–Saturday from 9:30 A.M. to 4:30 P.M. during the rest of the year.

Frederick Law Olmsted National Historic Site
99 Warren St., Brookline, MA 02146—617-566-1689

The designer of New York's Central Park, and founder of American landscape architecture, lived on this two-acre site during the last decade of his life. It is open Friday–Sunday from noon to 4:30 P.M.

Harvard University Museums of Natural History
Harvard University, Oxford & Divinity Sts., Cambridge, MA 02111—617-495-1910

The four museums in this group include the Geological and Mineralogical Museum. It has the oldest university mineralogy collection in the country. The Botanical Museum, home of the "Garden in Glass," displays over 700 models of plant species in completely life-like accuracy. The Museum of Comparative

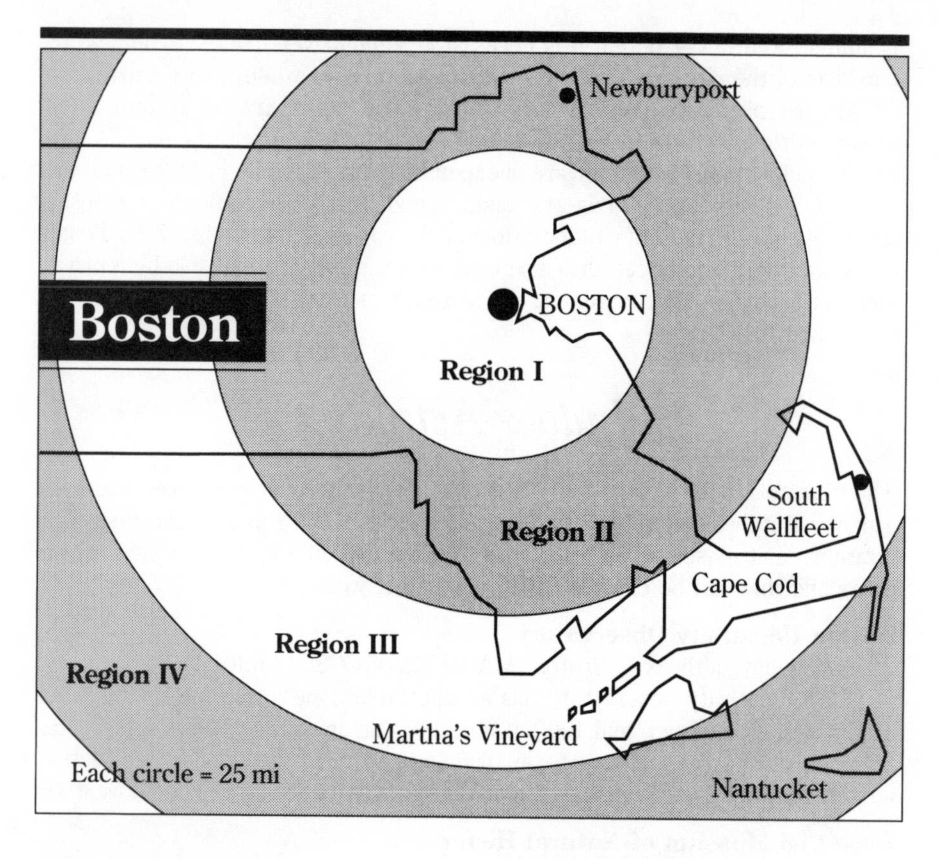

Indoor Activities

Region I

Boston University Observatory
Frederick Law Olmsted National Historic Site
Harvard University Museums of Natural History
Museum of Science
New England Aquarium

Region III

Cape Cod Museum of Natural History
National Marine Fisheries Service Aquarium

Region IV

Maria Mitchell Science Center

Outdoor Activities

Region I

Arnold Arboretum
Belle Isle Marsh Reservation
Blue Hills State Reservation
Boston Harbor Islands State Park
Broadmoor Wildlife Sanctuary
Charles River Reservation
Great Meadows National Wildlife Refuge
Ipswich River Wildlife Sanctuary
Long Hill Reservation
Walden Pond Reservation
Wompatuck State Park

Region II

Bolton Flats Wildlife Management Area
Parker River National Wildlife Refuge

Region III

Ashumet Holly Reservation and Wildlife Sanctuary
Wellfleet Bay Wildlife Sanctuary

Region IV

Felix Neck Wildlife Sanctuary

Zoology has exhibits from the earliest fossil forms to fish and reptiles of today. The museums are open Monday–Saturday, 9 A.M.–4:30 P.M. and Sunday 1 P.M.–4:30 P.M.

Maria Mitchell Science Center
1 Vestal St., Nantucket, MA 02554—508-228-2896
and
Museum of Natural Science
7 Milk St., Nantucket, MA 02554—508-228-0898

The Maria Mitchell Science Center is the restored home of America's first woman astronomer. It has herb and wildflower gardens in addition to information about Mitchell's works. The museum of Natural Science has exhibits of Nantucket wildflowers and birds. Both are open Tuesday–Saturday, 10 A.M.–4 P.M. from June 15th–August 31st.

Museum of Science
Science Park, Monsignor O'Brien Highway, Boston, MA 02114—617-723-2500

This science museum complex has a wide variety of exhibits from various fields of science such as astronomy, physical, and natural sciences. It includes the Hayden Planetarium. A number of exhibits feature New England ecology. It is open daily from 10:00 A.M. to 5:00 P.M.

National Marine Fisheries Service Aquarium
Water & Albatross Sts., Woods Hole, MA 02543—508-548-7684

This is a major research center for marine science located on Cape Cod. The aquarium contains numerous exhibits on local and commercial marine species, and a seal pool, as well as models of sea-going vessels. The aquarium is open daily during the summer from 10:00 A.M. to 4:30 P.M. During the rest of the year, it is open Monday–Friday, 9:00 A.M.–4:00 P.M.

New England Aquarium
Central Wharf, Boston, MA 02110—617-973-5200

North Atlantic ocean life is featured in this aquarium set on The Wharf along the shores of Boston Bay. It is open daily from 9:00 A.M. to 4:30 P.M.

Outdoor Activities

While it is pleasant whiling away a lazy winter or wet summer day picking up bits of information inside a nature museum, nothing is more invigorating than spending time outside. Experiencing nature firsthand by visiting a natural site, be it as small as a city lot or as large as the 28,000-acre Cape Cod National Seashore, is exhilarating. The following is a list of sites within the Boston–Cape Cod region.

Arnold Arboretum
125 Arborway, Jamaica Plain, MA 02130—617-524-1718

The gardens here contain the largest collection in the United States of woody trees and shrubs from the North Temperate Zone. The arboretum is a combined effort of Harvard University and the City of Boston. The goal of the first curator of the arboretum, Charles Sprague Sargent, was to find and grow every tree and shrub in the world that would survive Boston's winters. To that end over 7,000 types of trees and shrubs are grown. The arboretum is open daily from dawn to dusk.

Ashumet Holly Reservation and Wildlife Sanctuary
Ashumet Rd., Falmouth, MA 02540—508-563-6390

This 45-acre reserve located on Cape Cod began as a holly collection, and is now a bird-and-wildlife sanctuary with well-marked trails. It is open daily during daylight hours.

Belle Isle Marsh Reservation
146 Bennington St., Boston, MA 02128—617-727-5350

This 152-acre park in East Boston includes the last remnants of the salt marsh that once surrounded Boston Harbor. It is an excellent area for bird-watching.

Blue Hills State Reservation
695 Hillside St., Milton, MA 02186—617-698-5840

This 6,000-acre reserve is about ten miles south of Boston. It offers a wide range of native plants, birds, and small mammals, along with exposed granite domes called monadnocks. The highest of these rises over 600 feet to give climbers views of Boston and Boston Harbor. A network of trails crisscross the reserve and lead hikers to forested areas, rock outcroppings, meadows, large ponds, and a quaking peat bog. The park is frequently crowded because of its proximity to Boston.

The Massachusetts Audubon Society operates a trailside museum that is devoted to the natural history of the Blue Hills. It is open 10:00 A.M.–5:00 P.M. everyday except Monday.

Bolton Flats Wildlife Management Area
Headquarters, Division of Fisheries and Wildlife, 100 Cambridge St., Boston, MA 02202—617-727-3151
and
Oxbow National Wildlife Refuge
Headquarters, c/o Great Meadows NWR, 191 Sudbury Rd., Concord, MA 01742—617-369-5518

These wildlife refuges are 40 miles west of Boston. Take Interstate-495 to Exit 15. Follow State Route 117 to State Route 110. Go about 1 mile in either direction to the parking areas.

These two refuges total almost 1,500 acres, and offer an excellent birding area where egrets, ibises, and herons abound. Migration periods bring many ducks and geese. Small mammals are often seen, as well as white-tailed deer.

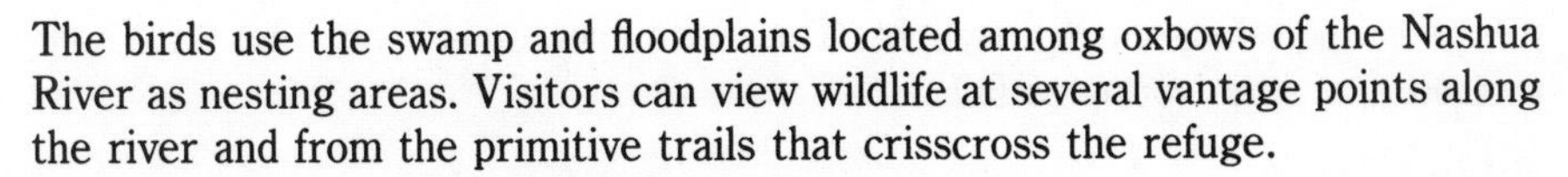
The birds use the swamp and floodplains located among oxbows of the Nashua River as nesting areas. Visitors can view wildlife at several vantage points along the river and from the primitive trails that crisscross the refuge.

Boston Harbor Islands State Park
Metropolitan District Commission, 20 Somerset St., Boston, MA 02108—617-727-5250

There are over 30 islands in Boston Harbor, all only minutes from downtown. Seven of these, Georges, Lovells, Peddock, Bumpkin, Great Brewster, Gallops, and Grape, form the Harbor Islands State Park. One other, Thompson, is privately owned, but open to the public on a limited basis.

The islands offer a variety of recreational activities, and all but Great Brewster and Thompson are served by water taxi or ferry. The other two are accessible by private boat. Most nature activities are found on Lovells, Gallops, Great Brewster, and Grape.

Broadmoor Wildlife Sanctuary
280 Eliot St., Natick, MA 01760—617-655-2296

This 577-acre sanctuary sits along the Charles River west of Boston, and has trails that wind in and around marshes, ponds, and meadows. A variety of wildlife can be seen here. The sanctuary is open daily from dawn to dusk. The visitor center is open Tuesday–Friday 9:00 A.M.–5:00 P.M. and Saturday–Sunday 10:00 A.M.–5:00 P.M.

Charles River Reservation
c/o Metropolitan District Commission, 20 Somerset St., Boston, MA 02108—617-727-5250

This 961-acre reservation extends along both sides of the Charles River from the dams at Boston and Cambridge to Newton Upper Falls. Naturalists will find little here other than a pleasant walk along the river where a number of birds may be spotted. The reserve is more for outdoor recreation such as swimming, tennis, golf, and picnicking.

Felix Neck Wildlife Sanctuary
Edgartown–Vineyard Haven Rd., Edgartown, MA 02539—508-627-4850

This 200-acre sanctuary includes beaches, marshes, open fields, and woodlands. Many waterfowl can be seen here, and the visitor center has exhibits, a waterfowl-rearing program, and a rehabilitation center for crippled birds. The sanctuary is open daily 8:00 A.M.–7:00 P.M. The visitor center hours are from 8:00 A.M. to 4:30 P.M.

Great Meadows National Wildlife Refuge
Headquarters, 191 Sudbury Rd., Concord, MA 01742—617-369-5518

The Concord unit of this refuge on Monsen Road is an excellent site for spring and fall bird-watching as it includes trails along river, marsh, and ponds. There are several interpretive trails in the unit.

The Sudbury unit on Weir Hill Road has nature trails along a river and through upland woods and swamp. The refuge visitor center is at this unit. The refuge is open daily during daylight. The visitor center is open from 8:00 A.M. to 4:00 P.M. from spring to fall.

Hiking Trails

Metropolitan District Commission, 20 Somerset St., Boston, MA 02108—617-727-5215

Although not all of the hiking trails maintained by the Metropolitan District Commission in the Greater Boston area are nature trails, they do offer an opportunity to walk and be outside. The MDC maintains trails in the Blue Hills Reservation in Milton; the Beaverbrook Reservation in Belmont; the Breakheart Reservation in Saugus; the Hemlock Gorge Reservation in Newton; and the Middlesex Fells Reservation and Belle Isle Marsh Reservation in Boston.

Ipswich River Wildlife Sanctuary

Perkins Row, Topsfield, MA 01983—508-887-9264

This reserve includes meadows, swamps, marshes, and ponds, as well as a stretch of the Ipswich River. This river snakes through the Great Wenham Swamp, where the 2,000-acre Ipswich River Wildlife Sanctuary, owned and operated by the Massachusetts Audubon Society, is located. The sanctuary offers a home to a variety of wildlife, with at least 130 species of birds, 25 of mammals, 18 of reptiles and amphibians, and 13 of fish. There are 15 miles of trails in the sanctuary, an observation tower overlooking the wildlife management area, and excellent bird-watching.

Many of the wildlife species can be observed on the guided canoe trips the Audubon Society leads through the sanctuary several times each year. For a $45.00 fee you can take a trip through one of the state's premier wildlife sanctuaries, complete with canoe, life jackets, full breakfast, and the director of the sanctuary as a guide.

On the trips you will explore the aquatic life of the river and swamp, glide through waterfowl feeding grounds, view a wide variety of bird life, and occasionally catch a glimpse of some of the small mammals such as otters and musk rats that reside in the sanctuary.

Novice canoeists are welcome on all these trips, but all participants must be able to swim. Contact the Ipswich River Wildlife Sanctuary for more information about the float trips and a trip application blank. In addition to the canoe trips, the Audubon Society has a number of wildlife and foliage walks at the sanctuary each year. Information about these can also be obtained from the Sanctuary.

Long Hill Reservation

Sedgewick Gardens, 572 Essex St., Beverly, MA 01915—617-922-1536

Located northeast of Boston on the Atlantic coast, this 14-acre reserve has fields, woods, and wetlands with more than 400 varieties of trees, shrubs, and flowers. Many of these are labeled. Nature trails cross the grounds of the estate. It is open daily from 8:00 A.M. to sunset.

Parker River National Wildlife Refuge
Northern Blvd., Plum Island, Newburyport, MA 01950—508-465-5753

The refuge is about 40 miles north of Boston and four miles southeast of Newburyport. This 4,662-acre refuge includes the southern two-thirds of Plum Island and the salt marshes west of the mainland. It has hiking trails, observation towers, and over six miles of beach. Bird-watching is excellent. The refuge is open daily from dawn to dusk.

Walden Pond Reservation
c/o Metropolitan District Commission, 20 Somerset St., Boston, MA 02108—617-727-5215

Thousands of tourists visit this 300-acre park each year to view the pond that Henry David Thoreau made famous. Unfortunately many have left litter and destruction. Still, you may wish to pay homage to one of America's great naturalists.

Wellfleet Bay Wildlife Sanctuary
P.O. Box 236, South Wellfleet, MA 02663—508-349-2615

Located on Cape Cod five miles south on U.S. Route 6, this 700-acre nature reserve has hiking trails, along with lectures, classes, and natural history field walks. It is open daily 8:00 A.M.–8:00 P.M. in summer, and from 8:00 A.M. to dusk Tuesday–Sunday during the rest of the year.

Wompatuck State Park
Union St., Hingham, MA 02043—617-749-7160

Located 20 miles southeast of Boston, take Exit 30 off State Route 3, and proceed three miles north on State Route 228.

This park overlooks Boston Harbor and has views of the Harbor Islands and Boston proper. Streams and small ponds dot the 2,900-acre park, and glacial material rises 50 to 100 feet above the surrounding countryside. The Forest Sanctuary Climax Grove, over 175 years old, has a wide variety of trees and shrubs that offer a habitat for birds and small mammals. The park is crowded during the summer. There are ten miles of designated hiking trails, a visitor center, and a nature study area.

Organizations That Lead Outings

One of the best ways to find out more about the natural history of a region is to join an outing led by a naturalist familiar with the flora and fauna. The following is a list of organizations in the Boston–Cape Cod region that lead such outings during the year. Contact those that have interests similar to yours to find out if they are offering any outings or lectures during your visit.

Charles River Canoe and Kayak Service
2401 Commonwealth Ave., Newton, MA 02166—617-965-5110

All boats other than inflatables are allowed on the Charles River and in the inner harbor. This rental service also offers some guided tours.

Friends of the Boston Harbor Islands
19 Myrtle St., Boston, MA 02114—617-523-8386
This organization is devoted to preserving the islands in the harbor. They occasionally lead tours and hikes.

Massachusetts Audubon Society
10 Juniper Rd., Belmont, MA 02178—617-489-5170

Nature Conservancy
294 Washington, Boston, MA 02139—617-423-2545

New England Aquarium
Central Wharf, Boston, MA 02110—617-973-5277

Sierra Club
3 Joy St., Boston, MA 02108—617-227-5339

Thompson Island Education Center
Thompson Island, Boston Harbor, Boston, MA 02109—617-328-3900
This center offers conference facilities and educational programs on privately owned Thompson Island. It is one of the largest of the over 30 islands in Boston Harbor.

Web of Life Outdoor Education Center
P.O. Box 530, Carver, MA 02330—508-866-5353

Wellfleet Bay Wildlife Sanctuary
P.O. Box 236, South Wellfleet, MA 02663—508-349-2615
Natural history field walks, lectures, and classes are conducted at the sanctuary. It is on Cape Cod.

Whale watching tours, with a naturalist on board, are offered by a number of commercial boats from Boston, Cape Cod, and Plymouth during the season, which normally runs from mid-April to mid-October. Some of those include the following.

Captain John Boat's Whale–Watch Tours
Town Wharf, Plymouth, MA 02360—508-746-2643

Hyannis Whale–Watcher Cruises
Barnstable Harbor, Barnstable, MA 02630—508-775-1622

New England Aquarium, Central Wharf, Boston, MA 02110—617-973-5277

Provincetown Whale–Watch Tours
MacMillan Wharf, Provincetown, MA 02657—508-487-2651

Web of Life Outdoor Education Center
P.O. Box 530, Carver, MA 02330—508-866-5353

A Special Outing

Most of Boston's residents and visitors drive about two hours for a chance to enjoy nature activities. Often they still find the crowds overwhelming. There are, however, beaches, saltwater marshes, dunes, and unspoiled wildlife areas within a short boat ride of downtown.

For those who venture seven miles across the harbor on one of the ferries serving George's Islands, the Boston Harbor Islands State Park offers abundant bird life, as well as many types of wildflowers. From George's Islands, water taxis carry visitors to the other islands in the park.

Sunbathers and swimmers favor the beaches of Lovells Island, but those who want more intimate contact with nature head for Peddocks, Bumpkin, and Grape Islands. Civilization has most encroached on Peddocks, where you can find running water, flush toilets, and cold drinks. But it also has a large saltwater marsh on West Head, which is one of four drumlins attached to the main island by narrow strips of land. This marsh is a major nesting ground for egrets, black-crowned herons, and cormorants.

Bumpkin and Grape Islands have few signs of civilization, and visitors must bring their own drinking water, as well as any other supplies they may need. They also must pack out all litter. These islands offer compensations for the lack of facilities, as visitors encounter lush fields with wildflowers such as Queen Ann's Lace and morning glories, and abundant bird life. Park guides offer interpretive nature walks during the summer, including some special ones for children. The islands offer an unusual site to visitors—wilderness with a view of Boston as a backdrop.

To find out more about the activities offered on the islands contact either the Metropolitan District Commission, Harbor Islands, 20 Somerset Street, Boston, MA 02108; 617-727-5250 or the Department of Environmental Management, Harbor Islands, 349 Lincoln Street, Building 43, Hingham, MA 02043; 617-740-1605.

George's Island is open from mid-April to Thanksgiving, while the others are open from mid-June through Labor Day. Contact the Thompson Island Education Center, Thompson Island, Boston Harbor, Boston, MA 02109; 617-328-3900, for information about the activities offered there.

Nature Information

Most parks in Boston focus on activities such as picnicking, golf, and tennis. Hiking trails are generally not emphasized even in those parks and preserves where nature is given notice. The following agencies have information about naturalist's activities in and around Boston.

American Youth Hostels, 1020 Commonwealth Ave., Boston, MA 02215—617-731-5430

Boston Department of Parks and Recreation
1 City Hall Plaza, Boston, MA 02108—617-725-4006

Bureau of Recreation
Division of Parks and Forests, Saltonstall Bldg., 100 Cambridge St., Boston, MA 02129—617-727-3180

Department of Commerce
Division of Tourism, Saltonstall Bldg., 100 Cambridge St., Boston, MA 02129—617-727-3201

Division of Fisheries and Wildlife
Saltonstall Bldg., 100 Cambridge St., Boston, MA 02129—617-366-4470
The Division of Fisheries and Wildlife publishes a booklet entitled *Wildlife Sanctuaries* that you may want to obtain.

Greater Boston Convention and Tourist's Bureau
Prudential Plaza, Boston, MA 02199—800-858-0200 or 617-536-4100

Massachusetts Department of Environmental Management
349 Lincoln St., Bldg. 43, Hingham, MA 02043—617-740-1605

Metropolitan District Commission
20 Somerset St., Boston, MA 02108—617-727-5250

Further Reading

AMC River Guide Committee. *AMC River Guide: Massachusetts, Connecticut, and Rhode Island.* Boston: Appalachian Mountain Club Books, 1985.

Berk, Susan, and Jill Bloom. *Uncommon Boston: A Guide to Hidden Spaces and Special Places.* Reading, MA: Addison-Wesley, 1987.

Brady, John, and Brian White. *Fifty Hikes from the Top of the Berkshires to the Tip of Cape Cod.* Woodstock, VT: Backcountry Publishers, 1983.

DeGraaf, Richard M., and Deborah D. Rudis. *Amphibians and Reptiles of New England: Habitats and Natural History.* Amherst, MA: University of Massachusetts Press, 1983.

———. *New England Wildlife: Habitat, Natural History, and Distribution.* U.S. Department of Agriculture, Forest Service, Northeastern Forest Experimental Station, General Technical Report NE-108. Washington, DC: U.S. Government Printing Office, 1986.

Dwelly, Marion. *Spring Wildflowers of New England.* Camden, ME: Down East Books, 1973.

Fisher, Alan. *AMC Guide to Country Walks Near Boston.* Boston: Appalachian Mountain Club, 1976.

Godin, Alfred J. *Wild Mammals of New England.* Chester, CT: Globe Pequot, 1983.

Hinds, Harold R., and W.A. Hathaway. *Wildflowers of Cape Cod.* Chatham, MA: Chatham Press, 1968.

Jorgenson, Neil. *A Sierra Club Naturalist's Guide: Southern New England.* San Francisco: Sierra Club Books, 1978.

Mallett, Sandy. *A Year With New England's Birds.* Somersworth, NH: New Hampshire Publishing Co., 1978.

Mitchell, J.H., and Whit Griswold. *Hiking Cape Cod.* Charlotte, NC: The East Woods Press, 1977.

Montgomery, Frederick Howard. *Seeds and Fruits of Plants of Eastern Canada and Northeastern United States.* Toronto: University of Toronto Press, 1977.

Primack, Mark L. *Greater Boston Park and Recreation Guide.* Chester, CT: Globe Pequot, 1983.

Sterling, Dorothy. *The Outer Lands: A Natural History Guide to Cape Cod, Martha's Vineyard, Nantucket, Block Island, and Long Island.* New York: Norton, 1978.

New York

New York's climate, human history, and geography are similar to those of Boston. As one of the largest cities in the United States, New York has over 7,000 acres of wetlands and urban wilderness within its city limits. When these wilderness sites are combined with the over 1,000 parks and playgrounds located in the city, the two total over 26,000 acres, or more than 17 percent of the total land area of the city. These parks and playgrounds contain almost three million trees.

In a time when the green and natural spaces of most cities are shrinking, New York's natural sites are flourishing. One sign of that is a move almost unheard of in these times of tight budgets. The New York City Department of Parks created a full-time staff of naturalists—the Natural Resource Group. During the last half of the 1980s this group mapped the vegetation and wildlife of the 7,000 acres of wilderness in the city, and developed a long-term plan that will protect

ABOVE: A great egret glides above a great blue heron in the Gateway National Recreation Area's Jamaica Bay Wildlife Refuge. The twin towers of Manhattan's World Trade Center loom in the background. National Park Service photo by Don Riepe.

these sites. They have also hired a full-time botanist, a person who was once arrested for collecting mushrooms and other wild, edible plants in Central Park, to lead wild food walks there, and in other city parks.

In addition to the acres of forested land in and around New York City, adjacent Long Island and New Jersey are blessed with several large, protected, wildlife refuges that serve as migration stopover points for numerous waterfowl and shorebirds. Other birds live in these refuges year-round.

Relatively uncrowded natural sites can be discovered within the city, and many others are within one to two hours of mid-town. A two-hour drive can take you to the middle of a wilderness swamp in New Jersey, the Catskills, or the outer reaches of Long Island. There, several wildlife refuges and open areas offer solitude where you can enjoy nature without the intrusion of honking cars, boom boxes, and flying objects such as Frisbees and footballs.

Climate and Weather Information

New York's climate is similar to Boston's, with cold, but moderate winters and hot, muggy summers. Highs in the mid-winter average near 40°F, and peak in the mid-80s in July and August. Fall and spring offer the most moderate temperatures for enjoying the outdoors.

The popular beaches and parks are filled to capacity by residents during the warm summer months. The traffic to these sites makes it difficult, at times, to reach relatively uncrowded nature sites that are located near them.

Getting Around New York

New York is another city where you do not want to drive unless you absolutely cannot avoid it. The streets are jammed, there is little parking, and visitors have a hard time finding unfamiliar locations.

Public transportation is excellent, reasonably priced, and easy to use. Many of the sites listed can be reached using public transit, but those outside the city are less likely to be. For outings to these sites you may wish to take public transportation to the outreaches of the city before renting a car, and return to your home base the same way.

Indoor Activities

New York has an abundance of museums and nature centers. You can visit them when you have only a short time to spend, or when the weather is so bad that you do not want to venture outside.

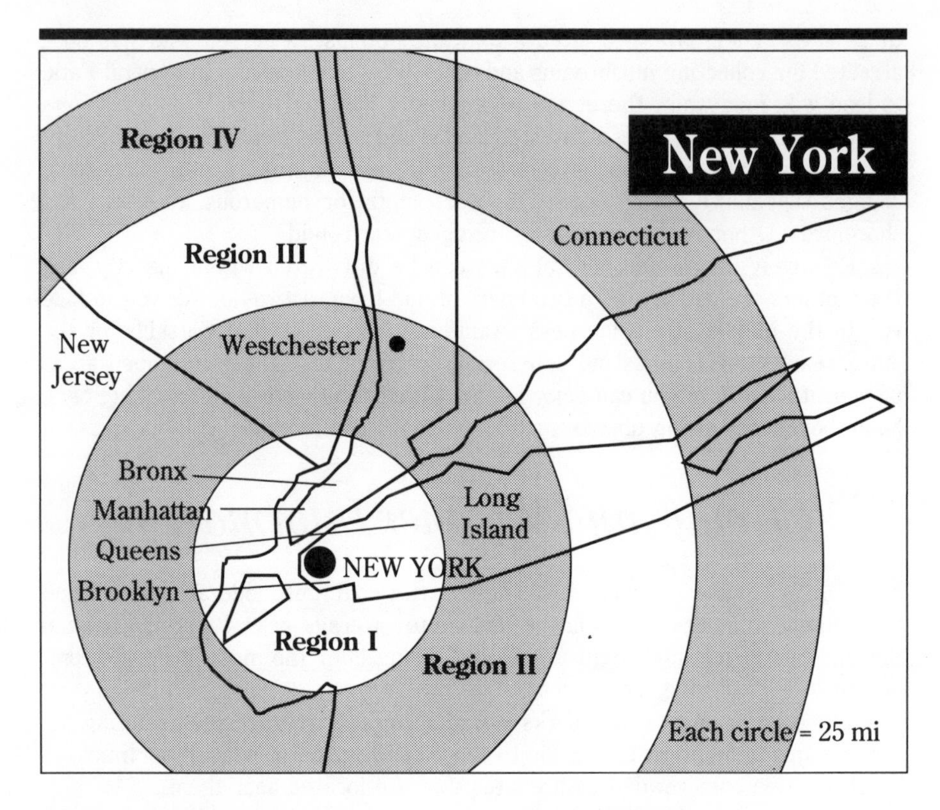

Indoor Activities

Region I

Alley Pond Environmental Center, Inc.
American Museum of Natural History
Brooklyn Botanic Garden
Brooklyn Children's Museum
Children's Museum of Manhattan
New York Aquarium
New York Botanical Garden
Prospect Park Environmental Center
Staten Island Museum

Outdoor Activities

Region I

Alley Pond Park
Brooklyn Botanic Garden
Central Park
Flushing Meadows-Corona Park
Gateway National Recreation Area
New York Botanical Garden
Pelham Bay Park
Prospect Park
Riverside Park
Seton Falls Park
The Greenbelt
Udall's Cove and Ravine Park

Region II

Arthur W. Butler Memorial Sanctuary
Caumsett State Park
Cheesequake State Park
Clark Garden
Planting Fields Arboretum
Roosevelt Bird Sanctuary and Trailside Museum
The Great Swamp

Region III

Bashakill Wetland
Connetquot River State Park Reserve

Region IV

Fire Island National Seashore
Morton National Wildlife Refuge

Alley Pond Environmental Center, Inc.
228-06 Northern Blvd., Douglaston, NY 11362—718-229-4000

The center offers exhibits of live animals and natural history of the New York area. It is open daily from 9:00 A.M. to 5:00 P.M.

American Museum of Natural History
Central Park West at 79th St., New York, NY 10013—212-769-5650

This is one of the foremost natural history museums in the nation. It has numerous exhibits on plant and animal life. The museum is open daily from 10:00 A.M. to 5:45 P.M.

Brooklyn Botanic Garden
1000 Washington Ave., Brooklyn, NY 11225—718-622-4433

This botanic garden has a subway station directly in front of it, one of the best Japanese gardens in the United States, a collection of flowering cherry trees second only to Washington, DC, a remarkable seasonal bloom, and a conservatory. All of this is on 50 acres next to the Brooklyn Museum and Prospect Park. While this is not a wilderness area, it does offer a respite from unbroken cement, and a place to go during unpleasant weather. The garden is open daily from 10:00 A.M. to 4:00 P.M.

Brooklyn Children's Museum
145 Brooklyn Ave., Brooklyn, NY 11213—718-735-4400

This museum frequently includes live animal exhibits, and has many hands-on activities. It is open 2:00 P.M.–5:00 P.M. daily and 10:00 A.M.–5:00 P.M. weekends and holidays.

Children's Museum of Manhattan
314 W. 54th St., New York, NY 10019—212-765-5904

This museum often has hands-on nature exhibits that feature live animals and the natural history of the area. It is open daily from 1:00 P.M. to 5:00 P.M.

New York Aquarium
Surf Ave. at West 8th St., Brooklyn, NY 11224—718-266-8711

Located on Coney Island, the aquarium has both outdoor pools and indoor tanks. It is open daily from 10:00 A.M. to 5:00 P.M.

New York Botanical Garden
Southern Blved. & 200th St., Bronx, NY 10458—212-220-8700

Located in northwest section of Bronx Park, this garden boasts one of the largest botanical collections in the world. There are over 250 acres of grounds, with a conservatory and museum building. The museum houses an herbarium with over four million plant specimens, and probably the best research library in the country. It is open daily from 10:00 A.M. to 4:00 P.M.

Prospect Park Environmental Center
1000 Washington Ave., Brooklyn, NY 11225—718-788-8548

This center, which sits near lakes and meadows, as well as near the Brooklyn Botanic Garden, offers classes, nature hikes, and lectures on the natural history of the Brooklyn area. It is open daily from 10:00 A.M. to 5:00 P.M.

Staten Island Museum
75 Stuyvesant Place, St. George, NY 10301—718-727-1135

This museum has extensive natural science collections that provide an excellent introduction to the natural history of Staten Island. It is open daily from 9:00 A.M. to 5:00 P.M.

Outdoor Activities

Outdoor naturalist's activities abound in and near New York. From wildlife refuges within the city to bird-watching in Brooklyn, visitors can enjoy a wide variety of experiences. Some choices are shown in the following list.

Alley Pond Park
228-06 Northern Blvd., Douglaston, NY 11362—718-229-4000

This 635-acre park in Queens has two major ecological sites. The first is a tidal basin where cord grass and salt shrub plant are part of an extensive salt marsh. The other is a forest upland located on a terminal moraine. The Alley Pond Environmental Center is located in this park.

Arthur W. Butler Memorial Sanctuary
and
Eugene and Agnes Meyer Nature Preserve
Chestnut Ridge Rd., Mt. Kisco, NY 10549—914-666-4221

The 350-acre Butler Sanctuary is owned and operated by the Nature Conservancy. It is an upland deciduous forest with many different habitats and 30 acres of swamp. There are over five miles of easy hiking trails in the sanctuary, and another five miles in the 35-acre Meyer Preserve. All of these interconnect to give visitors access to many scenic views and wildlife viewing areas. There is a visitor center at the sanctuary, and guided hikes are led in both the sanctuary and preserve year-round.

Bashakill Wetland
Wurtsboro, NY 12790—914-888-2711

On U.S. Route 209, this 2,200-acre wetland is one of the largest freshwater marshes in the state. It has thousands of different plants and animals, some of which are rare and endangered. The sanctuary area is off limits to visitors during some seasons. While there are few defined trails, there are some backroads and informal trails from which wildlife can be viewed.

Brooklyn Botanic Garden
1000 Washington Ave., Brooklyn, NY 11225—718-622-4433

See description in Indoor Activities section.

Caumsett State Park

RFD 3, Lloyd Harbor Rd., Huntington, NY 11743—516-423-1770

Located 35 miles east of New York City, this environmental park was once an estate. It offers ten miles of hiking trails. Over 150 species of birds have been recorded here, mostly waterfowl, shore and marsh birds. There are also many species seen in the oak-pine forests that cover about 60 percent of the 1,500 acres in the park.

Central Park

Central Park runs from 59th to 110th Street and from 5th Avenue to Central Park West. It is the first great urban park in the country, and one of the most famous. Central Park is 840 acres of land that contains a number of attractions. Some areas are so wild that people have actually built tree houses and lived in them for several years before being noticed.

Rangers conduct walking tours through the park, and "Wildman" Steve Brill, who was once arrested for collecting wild mushrooms and other wild edible plants from the park, leads wild food walks here and through other city parks. There are also a conservatory and zoo in the park.

Central Park sits in the center of New York City. It was first conceived in the 1840s when it became apparent that the lack of open green space in the city, and the lack of contact with nature, were becoming problems. People then were visiting the new cemeteries in Brooklyn and Queens to get a glimpse of nature.

In 1857 Calvert Vaux and Frederick Law Olmstead won the design competition for the planned park, and promised to "translate democratic ideas into trees and dirt." For over a hundred years Central Park has provided New York residents and visitors an island of nature in the city. It is an island where the strain of life indoors is relieved by a chance to be outdoors.

Unfortunately, some residents of the city have gravitated to the untamed natural sites of the park. This can make visits to the more isolated sections a little scary. With proper precautions and a congenial partner, however, these sections can be appreciated for their natural wildness.

The more popular sites in the park where you can view birds, observe spring wildflowers, and wander through forests with a wide variety of native and exotic trees and shrubs include: the Ramble, a 30-acre area with many pathways; the Reservoir, where joggers compete with waterfowl; and the Conservatory Garden, where formal gardens sit among the less formal areas of the park.

Springtime offers the most colorful and fragrant treats to park visitors, with cherries, magnolias, rhododendrons, and lilacs all adding their own special color and fragrance. Winter is also a special time as it gives visitors an opportunity to view the landscape through ice- and snow-covered trees devoid of leaves. Whatever the season, Central Park gives everyone an escape valve for the pent-up emotions that come from the pressures of urban living, where contact with nature often seems unattainable.

Cheesequake State Park
Matawan, NJ 07747—201-566-2161

This park is 30 miles south of New York City. Take the Garden State Parkway to Exit 120. Parking is less than one mile off the Parkway.

Cheesequake State Park has a wide array of habitat, including upland pitch-pine woods, mixed hardwood stands, freshwater marsh, salt marsh, and white-cedar bogs. It is one site where the pine-barren vegetation of southern New Jersey mingles with the deciduous forests of northern New Jersey.

The park is located on the inner coastal plain, which once was an agricultural area, but has become a heavy transportation and industrial zone. The woodlands in the park give a hint of the former rich diversity to be found on the inner plain.

A guided nature tour leads visitors through the diversity of the park, sometimes on a boardwalk. Spring brings many wildflowers to the woods, and they are at their best before the new leaves of the tree and shrub canopy cast a deep shade over the forest floor. The park is open daily from 8:00 A.M. to dusk. Some camping is available.

Clark Garden
193 I.U. Willetts Rd., Albertson, NY 11507—516-621-7568

This Long Island station of the Brooklyn Botanic Gardens has a 12-acre woodland and wildflower garden in addition to the more formal rose, vegetable, and herb gardens. Clark Garden is open Monday–Friday, 8:00 A.M.–4:30 P.M. and Saturday–Sunday, 10:00 A.M.–4:30 P.M.

Connetquot River State Park Reserve
State Route 27, Oakdale, NY 11769—516-581-1005

This 3,400-acre refuge on Long Island has guided nature walks and interpretive tours. Trails lead through wooded areas that support many species of mammals, birds, and plants. The preserve is open Tuesday–Sunday, 7:00 A.M.–4:30 P.M.. Call for nature walk times and reservations.

Fire Island National Seashore

Fire Island is off the south shore of Long Island, and has Robert Moses State Park and Smith Point County Park on either end of the National Seashore. Wildlife is abundant all along the shore, with waterfowl numerous between October and March, and mammals, such as whitetail deer, red foxes, and rabbits, often seen. A national wilderness area extends west for seven miles from Smith Point County Park.

There are a number of nature trails and visitor centers around the island, including a self-guided tour to the Sunken Forest at Sailors Haven, a self-guided nature trail at Watch Hill, and a visitor center at Smith Point County Park. Low shrubs and beach grass are found along the Atlantic shore, while high thickets and groves of pitch pines are more common in sheltered areas.

You can reach the island by car from the eastern end at Smith Point County Park. In the summer use the mainland ferry from Sayville and Patchogue.

Flushing Meadows-Corona Park
c/o New York City Department of Parks and Recreation, 64th St. & Fifth Ave., New York, NY 10021—212-360-1350

Bounded by Roosevelt Avenue, the Van Wyck Expressway, Union Turnpike, and 11th Street there are several regions that make up this park in Queens. The Willow Lake Nature Area has free urban ranger tours. Call 718-699-4204 for information on when these are held. Cattail Pond is a fresh water wetlands that supports a wide variety of aquatic insects and birds. Alley Pond Environmental Center often leads walks here.

Turtle Pond Interpretive Trail leads you by the site of a continental glacier, dense vegetation filled with wildlife, and kettle ponds. Call 718-229-4000 for information about guided tours. Forest Park is a natural, densely forested park with varied topography and abundant vegetation and wildlife. Guided tours with wild food collectors, naturalists, and bird-watchers are all frequently given.

Gateway National Recreation Area
Public Affairs Office, Bldg. 69 Headquarters, Floyd Bennett Field, Brooklyn, NY 11234—718-338-3687

Gateway, the fifth most visited site in the national park system, has four units. One is the Sandy Hook Unit in New Jersey, and the other three are in New York City. These are: Breezy Point, which includes the westernmost point of the peninsula; Jamaica Bay, which includes the Jamaica Bay Wildlife Refuge in Queens; and Staten Island, which includes Great Kills Park. A variety of recreational activities are found here, with naturalist's activities just one. Bird-watching and walks through marshland are both popular wildlife pursuits.

The Great Swamp National Wildlife Refuge
RD 1, P.O. Box 152, Basking Ridge, NJ 07920—201-647-1222

Take Interstate 287 south of Morristown to the Basking Ridge Exit. Take Lee's Hill Road and follow signs to refuge headquarters.

Although a bit farther by road, this refuge is only 26 air miles from Times Square. Plans for a New York City airport once threatened this swamp. Now some 3,600 acres of the 5,800-acre refuge are designated wilderness where migrant hawks and waterfowl rest on their journeys along the Atlantic Flyway.

The combination of thousands of plant species and outstanding wildlife makes this refugee a must-see. There are about five miles of unmaintained trails in the wilderness area, as well as the mile-long Laurel Trail at the Great Swamp Outdoor Education Center. A museum is also located at the education center, which is closed in July and August because of the number of insects in the surrounding swamp. The Wildlife Observation Center on the opposite side of the refuge has more than a mile of boardwalk trail through the swamp, interpretive displays, and a blind for observing wildlife close up.

The Greenbelt

One-hundred-year-old, second-growth woodlands and Richmond Creek are just two of the areas in the Greenbelt on Staten Island that offer sites seldom seen in urban areas. High Rock Park Conservation Center, a refuge that was rescued from a high-rise development, and Great Kills Park, with great fields of swamp grass, are two others. You can view Brooklyn and the World Trade Center from the latter.

Morton National Wildlife Refuge

c/o Wertheim National Wildlife Refuge, P.O. Box 21, Shirley, NY 11967—516-725-2270

This is a feeding and resting place on Long Island for migratory birds of the Atlantic Flyway. The piping plover and least tern are numerous here. There is a visitor center and self-guiding tour. Since the refuge is restricted to the public at certain times of the year, call to confirm open hours.

New York Botanical Garden

Southern Blvd. & 200th St., Bronx, NY 10458—212-220-8200

See description in Indoor Activities section. The Lorillard Waterfall on the Bronx River gives an excellent view of the many varieties of hardwood trees that live on the river's banks.

Pelham Bay Park

c/o New York City Department of Parks and Recreation, 64th St. & Fifth Ave., New York, NY 10021—212-360-1350

This is the largest public park in the Bronx, and is a good area to observe the geology of New York. Bedrock and glacial boulders here resemble the coast of Maine.

Planting Fields Arboretum

Mill River Rd. & Glen Cove Rd., Oyster Bay, NY 11771—516-922-9201

This 409-acre arboretum on Long Island includes landscaped greenhouses and gardens, plus some natural areas. Numerous species of trees can be found here, and many are labeled. The arboretum is open daily from 10:00 A.M. to 5:00 P.M.

Prospect Park

c/o New York City Department of Parks and Recreation, 64th St. & Fifth Ave., New York, NY 10021—212-360-1350

This 526-acre park has nature trails and a nature center. Guided nature tours are often held. This Brooklyn park is bounded by: Prospect Park West and Prospect Park Southwest, and Flatbush and Parkside Avenues.

Riverside Park

c/o New York City Department of Parks and Recreation, 64th St. & Fifth Ave., New York, NY 10021—212-360-1350

This four-mile park in Manhattan includes a bird sanctuary, as well as pleasant walks along the river. It is bordered by Riverside Drive, the Hudson River, 72nd Street, and 158th Street.

Roosevelt Bird Sanctuary and Trailside Museum
East Main St. at Cove Rd., Oyster Bay, NY 11771—516-922-3200

This memorial to President Roosevelt contains his grave and has a museum with examples of Long Island plant and animal life. Nature programs are presented on some weekends. The bird sanctuary is next to the museum and is open daily from 9:00 A.M. to 4:30 P.M.

Seton Falls Park
c/o New York City Department of Parks and Recreation, 64th St. & Fifth Ave., New York, NY 10021—212-360-1350

At Seton Avenue and East 233d Street in the Bronx, this is a natural urban park with diverse ecology including wooded hillsides, streams, marsh, and grassland. Hiking trails are featured throughout the park.

Udall's Cove and Ravine Park
c/o New York City Department of Parks and Recreation, 64th St. & Fifth Ave., New York, NY 10021—212-360-1350

This large salt marsh in Queens contains seawater from Little Neck Bay and Long Island Sound. It is diluted by freshwater from the surrounding uplands. Bird-watching is good here, and marsh life is easily observed. Hiking tails traverse the upland area of the park.

Organizations That Lead Outings

New York has numerous nature organizations that lead outings and sponsor lectures. Some of these are as follows.

Alley Pond Environmental Center
228-06 Northern Blvd., Douglaston, NY 11362—212-229-4000

Audubon Society
950 Third Avenue, New York, NY—212-832-3200

Beachfront New York
1422 Third Ave., New York, NY 10028—212-472-1031

Friends of Central Park
16 East 8th Ave., New York, NY 10014—212-473-7841

Look for Wild Foods
c/o New York City Department of Parks & Recreation, 64th St. & Fifth Ave., New York, NY 10021—718-291-6825

Outdoors Club
P.O. Box 227, Lenox Hill Station, New York, NY 10021—212-876-6688

Prospect Park Environmental Center
1000 Washington Ave., Brooklyn, NY 11225—718-788-8500

Shorewalkers (no mailing address)
212-663-2167

Sierra Club—Atlantic Chapter Outings
P.O. Box 880, New York, NY 10024—212-749-3740

Sierra Club New York City
625 Broadway, New York, NY 10012—212-473-7841

Urban Park Rangers
New York City Department of Parks and Recreation, 64th St. & Fifth Ave., New York, NY 10021—212-360-1350

Urban Trail Conference
P.O. Box 264, New York, NY 10274—718-720-1593

A Special Outing

Despite the fact that New York City is one of the world's largest metropolitan areas, and is known for its vast areas of unbroken asphalt and cement, it offers bird-watchers unlimited opportunities. Over 400 species of birds have been reported in the New York area, and nearly half of these have bred in the region during recent years.

The wide variety of habitat, from beach to inland woods that are found on hills with elevations up to 1,800 feet, are home to many nonmigrating and migrating species. Migrating waterfowl and shorebirds that use the Atlantic Flyway visit the area in large numbers during spring and fall.

Long Island is famous among bird-watchers for its rare birds, but its variety of common species makes it popular with casual observers as well. The region of Long Island most noted for its bird-watching is the western end where the Hudson Valley and the Atlantic coast join. Many migrating land birds pass by this junction during their voyages each year. Up to 38 species of warblers have been reported here in one season.

For those not familiar with the geography of Long Island, this prime bird-watching area is better known as Brooklyn. This is the very same borough of New York that is noted for its vast neighborhoods of brownstones and tenements, where residents are thought to dream of growing even a small tree. There are numerous areas in Brooklyn that defy that vision, however, and offer many sites of birds nesting and feeding.

The favorite bird-watching sites in the borough are: Prospect Park, where the Rose Garden, Vale of Cashmere, and shores of Prospect Lake are popular; Marine Park; the marsh on the south side of Avenue U; and the Jamaica Bay

Wildlife Refuge, which is east of Prospect Park, and completely within New York City limits. Riis Park, part of the Gateway National Recreation Area, is a quarter-hour drive west of Jamaica Bay, and is another excellent site for the fall migration.

So when you think of Brooklyn, think of bird-watching. It is as good there as you will find it in or near any major population center in the country.

Nature Information

As can be expected in a city where nature has been given priority, information about nature sites and activities is fairly easy to obtain. The following is a list of some of the governmental agencies that will provide you with current information about what is occurring at different sites around the city.

New York Office of Parks and Recreation
Agency Bldg. No. 1, Empire State Plaza, Albany, NY 12238—518-474-0456

New York Convention and Visitors Bureau
2 Columbus Circle, New York, NY 10019—212-397-8200

Division of Fish and Wildlife
State Environmental Conservation Department, 50 Wolf Rd., Albany, NY 12233—518-457-3522
Various publications can be ordered from the State Environmental Conservation Department, Publications, Room 111. Among these are the *Guide to Outdoor Recreation in New York State and Hiking Areas of New York State.* A publications list is available on request.

New York City Department of Parks and Recreation
64th St. & Fifth Ave., New York, NY 10021—212-360-1350

Write or call to receive a copy of brochure entitled *The Green Pages: A Guide to New York City Parks.*

Further Reading

Albright, Rodney, and Priscilla Albright. *Short Walks on Long Island.* Chester, CT: Globe Pequot, 1983.

Arbib, Robert S., Jr., Olin Sewall Pettinggill, Jr., and Sally Hoyt Spofford. *Enjoying Birds Around New York City.* Boston: Houghton Mifflin, 1966.

Bennet, D.W. *New Jersey Coast Walks.* Highlands, NJ: American Littoral Society, 1981.

Berran, Paul, and David F. Karnosky. *Street Trees for Metropolitan New York.* No. 1. Poughkeepsie, NY: New York Botanical Garden Institute of Urban Horticulture, 1983.

Boyle, William J., Jr. *New Jersey Field Trip Guide.* Summit, NJ: Summit Nature Club, 1979.

Bull, John. *Birds of the New York Area.* New York: Dover, 1964.

Connor, Paul F. *The Mammals of Long Island, New York.* Bulletin 416. Albany, NY: Bulletin 416, New York State Museum and Science Service, 1971.

Cornell Laboratory of Ornithology. *Enjoying Birds Around New York City.* New York: Houghton Mifflin, 1966.

Gardner, Jean. *Urban Wilderness: Nature in New York City.* New York: Earth Environmental Group, 1988.

Graaf, M.M. *Tree Trails in Central Park.* New York: Greensward Foundation, 1970.

Kieran, John. *A Natural History of New York City.* Boston: Houghton Mifflin, 1959.

Knowler, Donald. *The Falconer of Central Park.* New York: Bantam, 1984.

Leck, Charles. *The Birds of New Jersey: Their Habits and Habitats.* New Brunswick, NJ: Rutgers University Press, 1975.

New York-New Jersey Trail Conference. *New York Walk Book.* New York: Doubleday, 1971.

Ricciuti, Edward R. *The New York City Wildlife Guide.* New York: Nick Lyons Books, 1984.

Scheller, William G. *Country Walks Near New York.* Boston: Appalachian Mountain Club, 1980.

Schuberth, Christopher J. *The Geology of New York and Environs.* New York: Natural History Press, 1968.

Thomas, Bill and Phyllis. *Natural New York.* New York: Holt, Rinehart & Winston, 1983.

Philadelphia

Philadelphia is as rich in early American historical lore as Boston, and developed contemporaneously with New England. Unlike Boston it retained more of its original natural environment as it grew into a major metropolitan area.

The original forests of Pennsylvania were magnificent stands of hardwood trees, shrubs, and undergrowth. They were home to a wide variety of wildlife. Today's remnants of those forests lack the magnificence of the original stands, but they make the exurban region of Philadelphia one of the most luxurious of any American city. Bucks, Chester, Delaware, and Montgomery Counties, all within easy commuting distance of Philadelphia, have large tracts of rolling hills and woodlands that offer a pastoral haven to Philadelphia's residents and visitors.

Some Pennsylvania forests are home to species of wildlife that are more abundant today than they were in 1900. Streams run clear in these forests. The

---

ABOVE: Visitors can observe waterfowl on ponds in the Tinicum National Wildlife Refuge, and learn about them at the Tinicum National Environmental Center near downtown Philadelphia. Photo courtesy of U.S. Fish and Wildlife Service.

two million acres of state forests have more than 2,500 miles of marked trails that are used by outdoor enthusiasts from around the country.

While Philadelphia does not sit on a large bay or along a coast, it is close enough that national wildlife refuges of the New Jersey shores are only about an hour away. This allows visitors to explore the forested regions of southeast Pennsylvania, the riparian life along the banks of the Schuylkill and Delaware Rivers, the pine barrens of southern New Jersey, and the salt marshes of the New Jersey coast. Huge flocks of migratory birds using the Atlantic flyway rest and feed during the fall and spring on the coast. All of this is less than two hours from downtown Philadelphia.

Climate and Weather Information

Philadelphia has the hot, humid summers and relatively mild winters that are common to the mid-Atlantic region. Winters are mild enough for some outdoor activities most of the season. Fall and spring offer moderate times when the New Jersey shore and pine barrens are pleasant. The upland woods give visitors some respite from the heat and humidity of mid-summer. Be prepared for rain during the spring and summer.

Getting Around Philadelphia

Philadelphia has an extensive public transportation system, but few of the outdoor naturalist outings listed are served by it. The indoor activities, including museums, arboretums, conservatories, and nature centers, are generally accessible using public transportation.

Downtown Philadelphia has many old, narrow streets that make driving difficult, and parking is minimal. Taxis are plentiful, but fares are not the cheapest. Rental cars agencies can be found at convenient locations throughout the city.

Indoor Activities

The following activities can be enjoyed any time of the year, but are especially appropriate during times when winter cold drives you inside, or the heat and humidity makes the outside unbearable. At various times, several city universities, such as the University of Pennsylvania, St. Josephs University, and Temple University, have natural science exhibits in their museums. Such exhibits are normally listed in local newspapers.

Academy of Natural Sciences
19th St. & Benjamin Franklin Pkwy., Philadelphia, PA 19103—215-299-1000

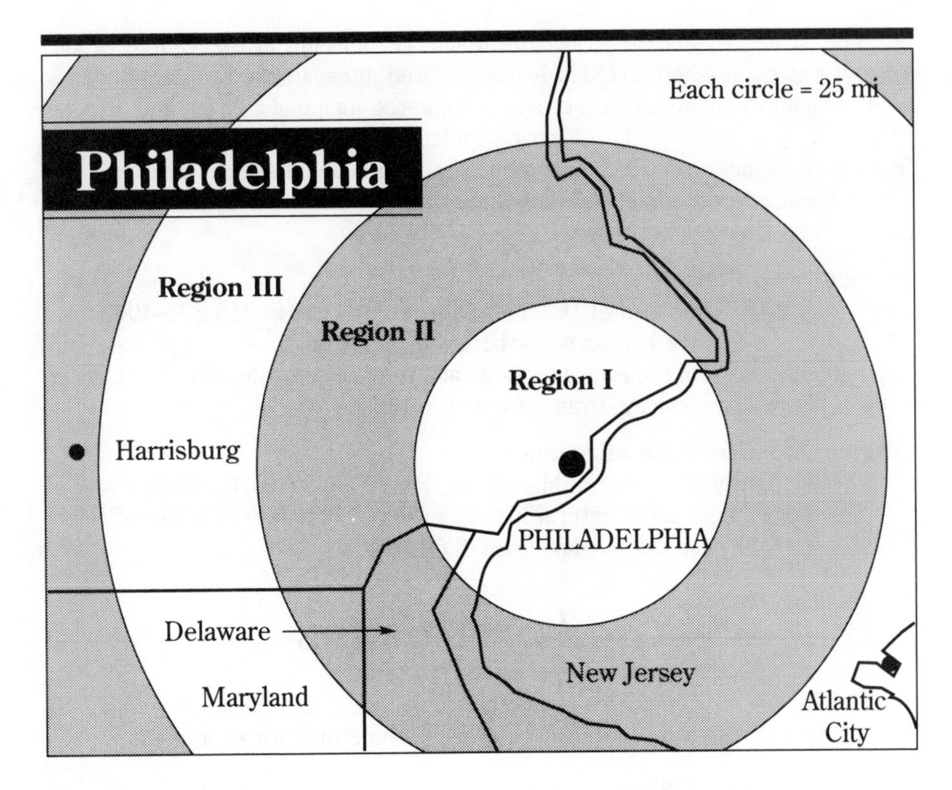

Indoor Activities

Region I

Academy of Natural Sciences
Horticultural Center
Wagner Free Institute of Science

Outdoor Activities

Region I

Andorra Natural Area
Bartram's Garden
Carpenter's Woods
Fairmount Park
Morris Arboretum
Ridley Creek State Park
Schuylkill Valley Nature Center
Tinicum National Environmental Center
Tyler Arboretum

Region II

Audubon Wildlife Sanctuary
Bowman's Hill State Wildflower Preserve
French Creek State Park
New Jersey Pine Barrens
Nockamixon State Park
Wharton State Forest

Region III

Edwin B. Forsythe Wildlife Refuge
Hawk Mountain Sanctuary
Middle Creek Wildlife Management Area

The Academy was the first natural science museum in the United States when it was established in 1812. It has achieved international reputation for its natural habitat exhibits of extinct and endangered mammals, birds, and insects. "Discovering Dinosaurs" is its most popular exhibit. This multi-media exhibit features Tyrannosaurus Rex. Live animal programs and a special children's nature museum are also popular. The academy is open daily year-round from 10:00 A.M. to 5:00 P.M.

Horticultural Center
Belmont Ave. & Horticultural Dr., Philadelphia, PA 19104—215-879-4062

There is a 28-acre landscaped arboretum, Japanese house and garden, and large greenhouse with thousands of plants and trees used by the city. The Center is open Wednesday–Sunday, from 9:00 A.M. to 3:00 P.M.

Wagner Free Institute of Science
17th St. & Montgomery Ave., Philadelphia, PA 19126—215-763-6529

This museum features natural history exhibits. It is open Tuesday–Friday, 10:00 A.M.–4:00 P.M. and Sunday, noon–3:00 P.M.

Outdoor Activities

Philadelphia has a wide choice of outdoor activities. From upland woodlands to pine barrens to seashore marshlands, there is something for everyone.

Andorra Natural Area
Northwestern Ave., Philadelphia, PA, 19118—215-242-5610

This National Natural Landmark in the Wissahickon Valley is only minutes from downtown. Its steep slopes are covered with hardwood forests, and its streams run full with water fresh enough to support trout. Thousands of migrating birds gather there in fall and spring. The geological history of the region can be read from the outcroppings of ancient rocks. Deer, skunks, raccoons, and red foxes are abundant in the 2,000-acre park.

Audubon Wildlife Sanctuary
Audubon, PA 19407—215-666-5593

Mill Grove, the first American home of John James Audubon, is located on this 130-acre sanctuary. The mansion serves as a museum, and the rest of the grounds are in a natural state except for the six miles of trails. The forest and creek sides are inhabited by at least 400 species of flowering plants and 175 species of birds. The sanctuary is open daily from 10:00 A.M. to 5:00 P.M., except Monday.

Bartram's Garden
54th St. & Lindberg Blvd., Philadelphia, PA 19143—215-729-5281

George Washington, Benjamin Franklin, and Thomas Jefferson all visited this 27-acre garden on the banks of the Schuylkill River. It has changed little since their time. Claimed as America's first botanical garden, it was founded in 1728 by John Bartram, who was the Royal Botanist to King George III. The gardens are open daily from dawn to dusk. A restored farmhouse is open 10:00 A.M.–4:00 P.M., Tuesday–Friday.

Bowman's Hill State Wildflower Preserve

Washington Crossing Historic Park, Route 32, New Hope, PA 18938—215-493-4076

Located three miles south of New Hope, this preserve is a living natural museum of the native plants of Pennsylvania. Anyone interested in native plants will find this trip outside of Philadelphia well worth the time. Songbirds are plentiful here during the spring migration, and the top of Bowman's Hill is an excellent spot for observing treetop migrants.

Carpenter's Woods

c/o Andorra Natural Area, Northwestern Ave., Philadelphia, PA 19118—215-242-5610

Carpenter's Woods is on Wayne Avenue off Wissahickon Avenue. This valley, covered with mature forest, is one of the best birding spots in Philadelphia. The most spectacular collection of migratory birds comes in May, but August and September also offer excellent showcases of migrating land birds. Hiking trails crisscross the 40 acres in the park.

Edwin B. Forsythe National Wildlife Refuge

P.O. Box 72, Oceanville, NJ 08231—609-652-1665

This refuge is north of Atlantic City on State Route 9, about one and one-half hours from Philadelphia. More than 34,000 acres of bays, salt marshes, barrier beaches, dunes, upland fields, and woodlands make up this popular refuge. In 1984, when it was renamed to honor the late Congressman Forsythe, this refuge combined the Brigantine and Barnegat Wildlife Refuges. A major stop on the Atlantic Flyway, the peak migrations seasons are from mid-March to mid-April and from mid-October to mid-December.

Fairmount Park

25th St. & Franklin Pkwy., Philadelphia, PA 19133—215-686-2176

This 8,500-acre park is the largest landscaped city park in the world. It is Philadelphia's answer to Central Park. Sports facilities, museums, playgrounds, and all the other attractions of a city park often distract visitors from the woods, meadows, and other natural sites along both sides of the Schuylkill River.

French Creek State Park

R.D. 1, Elverson, PA 19520—215-582-1514

Located on Pennsylvania Turnpike between Philadelphia and Reading, this 6,841-acre park is in the midst of the most densely populated section of state. It is heavily used during the summer. In the park, there is a blending of northern and southern hardwoods as well as many flowering plants. Its two lakes attract waterfowl such as herons. There is a nature center and an interpretive program, along with hiking trails. These include eight miles of the 120-mile Horseshoe Trail that extends from Valley Forge to the Appalachian Trail.

Hawk Mountain Sanctuary
Route 2, Kempton, PA 19529—215-756-6961

About one hour northwest of Philadelphia, this sanctuary is near Hamburg, Pennsylvania. This is one of the few sanctuaries in the world set aside for migrating birds of prey. An average of over 15,000 migrating raptors are estimated to pass through the sanctuary in a season. More than 200 species of other birds join the 14 species of raptors on their flights along the migration route. The greatest number of raptors are seen from late August to the end of November. The area is noted for its rhododendron and mountain laurel bloom in May and June. The museum has birds of prey exhibits and observation windows. It is open daily from 8:00 A.M. to 5:00 P.M.

Middle Creek Wildlife Management Area
R.D. 1, Newmanstown, PA 17073—717-733-1512

The first 1,700 of the 5,000 acres at the site near Lancaster were acquired during the 1930s. The rest were acquired during the 1960s and 1970s to develop waterfowl habitats. A 400-acre shallow lake, and an additional 70 acres of ponds and potholes were created to develop the habitats. Nine miles of trails are within the site, plus an observation point, waterfowl habitat view, and visitor center. From March 1st to November 30th, the visitor center is open Wednesday–Saturday, 10:00 A.M.–5:00 P.M. and Sunday, 1:00 P.M.–5:00 P.M.

Morris Arboretum
9414 Meadowbrook Ave., Philadelphia, PA 19118—215-247-5777

This 175-acre arboretum is located about 15 miles from downtown Philadelphia in suburban Chestnut Hill. This is not the best place to study native trees of the region except on the woodland trails near Wissahickon Creek. Exotic woody plants, particularly from China and the Far East, are emphasized. The arboretum is open daily from 9:00 A.M. to 4:00 P.M.

New Jersey Pine Barrens
c/o Wharton State Forest, Batsto, R.D. 1, Hammonton, NJ 08037—609-561-3262

One of the most unexpected wilderness experiences in the East is found about an hour's drive east of Philadelphia in New Jersey. About 1.6 million acres of pine barrens are sandwiched between the interior piedmont of Pennsylvania and the coast of New Jersey.

This region has abundant small bird and mammal populations, as well as numerous forms of unusual plant life. All that live and grow in stands of pine can be found on wide expanses of white sand. Natural bogs, swampland, and wild streams are interspersed with the forests, where it is possible to hike or canoe for a full day without encountering another person.

There is only one long, established trail in the Barrens. It is the 30-mile Batona Trail, located in the 97,578 acres of the Wharton State Forest in the center of the Barrens. There are several short nature trails at park headquarters at Batsto that introduce visitors to the botanical growth common to the region.

The Batona Trail is one of the finest wilderness trails in the East. It is an excellent trail for all ages and offers no hardships or obstacles along its length. The trail follows along the banks of Batsto River for the first eight miles, and crosses a diversity of terrain the rest of the way. Wildflowers are in abundance during the spring, when pitcher plants, sundew, turkey beard, sheep laurel, and pixie moss are all evident. Wild blueberries and huckleberries can be found by the handful later in the season.

Bird-watching is excellent in the forest, along the Batona Trail, and along the over 400 miles of sand roads that are virtually traffic free, particularly on weekdays. While it can be hot in the Barrens during the summer and fall, it is generally dry, with cool nights. Winter days are mild, and hikers are even less likely to meet others then.

For visitors who wish to camp in the state forest there are nine campgrounds, of which only one has organized sites. Water from hand pumps is available at most of the camps. Permits must be obtained from the forest headquarters. You can get more information about Wharton State Forest from the superintendent at the address listed.

Nockamixon State Park

R.D. 3, Quakertown, PA 18951—215-257-3646

This 5,000-acre day-use park on State Route 563 is crowded during the summer months. Over 300 species of wildflowers and 190 species of birds have been identified here. Most trails in the park are short, but some extend into over 2,000 acres of state game land adjacent to the park. Interpretive programs are available during the summer.

Ridley Creek State Park

Route 6, Sycamore Mills Rd., Media, PA 19063—215-566-4800

This park is southwest of Philadelphia on State Route 3 near Edgemont. Woodlands and meadows bisected by Ridley Creek make up this 2,390-acre park which includes over ten miles of hiking trails. Migrating songbirds are plentiful here and in adjoining Tyler Arboretum.

Schuylkill Valley Nature Center

8480 Hagy's Mill Rd., Roxborough, PA 19128—215-482-7300

This center has 500 acres of fields, thickets, ponds, streams, and woodlands. A true wildlife haven in the city, it offers seven miles of trails that wind through nature areas. It is open Monday–Saturday, 8:30 A.M.–5:00 P.M.

Tinicum National Environmental Center

86th St. & Lindberg Blvd., Philadelphia, PA 19153—215-521-0662

This center is the largest remaining wetland in Pennsylvania, and one of the few wildlife refuges to be found entirely within the city limits of a major urban area. The others are in New York and Minneapolis. This center includes 1,200 acres of freshwater marsh, tidal creek, and second growth forests. Thousands of ducks gather here in fall and winter, and there is a rookery of herons and egrets each summer. Over 280 species of birds have been sited in the refuge. It is an island of greenery in one of the most heavily developed portions of the Delaware River, but is also one of the main passageways for birds along the Atlantic Flyway. An observation tower, visitor center, walking trails, and a boardwalk all add to the views of birds and wildlife. The refuge is open daily from 8:00 A.M. to sunset. The visitor center is open from 8:30 A.M. to 4:00 P.M.

Tyler Arboretum
P.O. Box 216, 515 Painter Rd., Lima, PA 19060—215-566-9133

Over 20 miles of trails lead visitors throughout the 700 acres in this collection of trees and shrubs. It is located about 13 miles southwest of downtown Philadelphia. A combination of native and introduced species cover wooded slopes with mature forests, a stream valley, and the one remaining serpentine barren in Delaware County. Spring and fall songbird migrations are outstanding here. It is open daily from 8:00 A.M. to 5:00 P.M., with extended summer hours. Ridley Creek State Park adjoins the arboretum grounds.

Wharton State Forest
Batso, R.D. 1, Hammonton, NJ 08037—609-561-3262

This forest is on Route 542 off the Garden State Parkway about one hour east of Philadelphia. Two rivers, one lake, numerous streams, a nature trail, and the 40-mile Batona Wilderness Trail are all attractions to this large pine forest in New Jersey. There are also over 500 miles of sandy roads in the forest that are open for hiking. Atison Lake is crowded during summer weekends, but most of the rest of the forest offers a serene escape from urban noise.

Organizations That Lead Outings

Academy of Natural Sciences
19th St. & Benjamin Franklin Pkwy., Philadelphia, PA 19103—215-299-1000

Andorra Natural Area
Northwestern Ave., Philadelphia, PA 19118—215-242-5610

Schuylkill Valley Nature Center
8480 Hagy's Mill Rd., Roxborough, PA 19128—215-482-7300

A Special Outing

Wissahickon Creek, which runs through Fairmount Park, has cut through outcroppings of rock to reveal the complex geological history of the Philadelphia region. Just minutes northwest of downtown, this National Natural Landmark provides hikers with opportunities to view various forms of igneous and metamorphic rock that owe their presence to the building of the Appalachian Mountains to the east.

The following rocks are all exposed along the banks of the creek, and can be examined as you walk along an unmarked, mile-long trail: Baltimore gneiss, a pink-banded, metamorphic rock that is at least one billion years old and is the oldest formed in the region; serpentine, a metamorphic rock formed as tectonic plates ground together in the process that uplifted the Appalachian Mountains; and schist, formed from shale and mudstone as they were subjected to intense heat and pressure beneath the earth's surface. Rounded granite boulders, igneous rocks, can be seen in the stream bed where erosion has rounded them. They fell from an outcropping that can be seen to the north.

The Department of General Services, State Book Store, P.O. Box 1365, Harrisburg, PA 17125, has copies of the *Guidebook to the Geology of the Philadelphia Area* (Bulletin G 41). It describes four geology walks in the Wissahickon Valley. Complete directions to this walk, and other natural sites in the Wissahickon Valley can be obtained from the Andorra Natural Area, Northwestern Avenue, Philadelphia, PA 19118; 215-242-5610.

Nature Information

Agencies that provide information on naturalist's activities in the Philadelphia region include the following.

Andorra Natural Area
Northwestern Ave., Philadelphia, PA 19118—215-242-5610

Bucks County Tourist Commission
152 Swamp Rd., Doylestown, PA 18901—215-345-4552

Bureau of Forestry
Department of Environmental Resources, P.O. Box 1467, Harrisburg, PA 17120—717-787-2708

Bureau of State Parks
Department of Environmental Resources, P.O. Box 1467, Harrisburg, PA 17120—717-787-6640

Department of General Services
State Book Store, P.O. Box 1365, Harrisburg, PA 17125

Philadelphia Convention and Visitors Bureau
1525 John F. Kennedy Blvd., Philadelphia, PA 19102—215-636-1666

Further Reading

Bennett, D.W. *New Jersey Coast Walks.* Highlands, NJ: American Littoral Society, 1981.

Boyle, William J., Jr. *New Jersey Field Trip Guide.* Summit, NJ: Summit Nature Club, 1979.

Fisher, Alan. *Country Walks Near Philadelphia.* Boston: Appalachian Mountain Club, 1982.

Geyer, Alan R., and William H. Bolles. *Outstanding Geological Features of Pennsylvania.* Harrisburg, PA: Bureau of Topographic and Geologic Survey, 1979.

Goodwin, Bruce K. *Guidebook to the Geology of the Philadelphia Area.* Harrisburg, PA: Bureau of Topographic and Geologic Survey, 1964.

Hoffman, Carolyn. *Fifty Hikes in Eastern Pennsylvania.* Woodstock, VT: Backcountry Publications, 1982.

Klimas, John E., Jr. *The Wild Flowers of Pennsylvania.* New York: Walker & Co., 1975.

Washington, DC

The District of Columbia was built amidst unbroken swamp and natural woodlands. Since its beginning in 1791, green parks and woodlands have been part of its ambience. Today the city has one of the highest ratios of parks to residents of any urban area in the United States. Scenic woodland trails crop up in the

ABOVE: The Jug Bay Natural Area in Maryland's Patuxent River Park offers solitude to visitors during late winter. Photo courtesy of Maryland–National Capital Park and Planning Commission.

most unlikely places in the city, and the noise of rush hour traffic often disappears as you enter a quiet pathway where time seems to have stood still.

Virgin timber, native wildflowers, and literally hundreds of varieties of trees and shrubs all stand on large natural spaces. Many are accessible from the busy streets and freeways of downtown. Other natural areas, from mountain parks to salt marshes along the Chesapeake Bay, are less than an hour's drive from Capitol Hill.

The Appalachian Mountains, and the Appalachian Trail, are only 50 miles west of the city, and the Atlantic Ocean is 130 miles east. Between the two lies the Chesapeake Bay, and two of its feeder rivers, the Potomac and the Anacostia. The Potomac is literally the front door to the city, while the Anacostia cuts through its eastern sections.

Although Washington, DC does not sit on the coast as New York and Boston do, water still plays an important role in the naturalist's activities available. Rock Creek Park, Anacostia Park, Great Falls Park, and the Chesapeake and Ohio Canal National Historical Park all border the rivers. Many favorite bird-watching sites of the region are along arms of the Chesapeake Bay.

About the only life zone common to the Northeast not within easy reach of the city, is that of the seacoast where sand dunes and ocean beach dominate. The areas of interest discussed in this section include a variety of riparian habitats, upland forests, salt and freshwater marshes, and outcroppings of rock.

Climate and Weather Information

Washington has many of the same climatic conditions as Boston, New York, and Philadelphia, but is somewhat warmer during the summer and winter. Snow still falls during deep winter since the ocean is too far away to be a moderating influence. Many late fall, winter, and early spring days are very conducive to outdoor activities.

Summer can be hot and humid, but you can find outings to the east in the foothills of the Appalachians that are more moderate. Spring is undoubtedly the most beautiful season as the wildflowers of the region carpet the forests and meadows. Fall offers cool, dry days, and migrating birds are plentiful. Rain is prevalent during spring and summer.

Getting Around Washington

The city public transit system is good, and many of the small parks and gardens can be reached using it. Many sections of the extensive Rock Creek Park are also served, as well as part of the Chesapeake and Ohio Canal. Taxis are frequent, and fares average. Washington is a little easier to drive in than the other cities in the Northeast, but you may still want to rent a car on the outskirts to visit outlying areas.

Indoor Activities

Washington is famous for its national museums, and several of these are devoted to natural history. The following list includes some indoor treats when you do not feel like challenging the outdoors.

Maryland Science Center
601 Light St., Baltimore, MD 21230—301-685-5225

The science center has extensive exhibits on the ecology and wildlife of the Chesapeake Bay region. It is open daily from 10:00 A.M. to 8:00 P.M.

National Aquarium
14th St. & Constitution Ave., NW, Lower level of the Commerce Bldg., Washington, DC 20006—202-377-2825

This is the oldest public aquarium in the United States, and displays more than 1,000 fish and sea creatures representing over 200 species. It also has a touch tank. The aquarium is open daily from 9:00 A.M. to 5:00 P.M.

National Aquarium in Baltimore
501 Pratt St., Baltimore, MD 21202—301-576-3810

This new, seven-level modern structure includes three major areas: an Atlantic Coral Reef, a Tropical Rain Forest, and Man and the Sea exhibit. The aquarium is open Monday–Thursday, 9:00 A.M.–5:00 P.M. and Friday–Saturday, 9:00 A.M.–8:00 P.M.

National Museum of Natural History of the Smithsonian Institution
10th St. & Constitution Ave., NW, Washington, DC 20560—202-357-2700

This museum is unmatched for the number and depth of its exhibits. From the children's Discovery Room to the Splendors of Nature exhibit there is something here for all interests. It is open daily from 10:00 A.M. to 5:30 P.M.

Outdoor Activities

In addition to its many parks, Washington is known for its extensive gardens. Many of these are included in the following list, and you are likely to find them good bird-watching areas. All of the counties in Virginia and Maryland surrounding Washington have parks with good hiking trails. Most of these are included in this list, but you may find more in your travels around the city.

Anacostia Park
c/o National Park Service, 1100 Ohio Dr., SW, Washington, DC 20242—202-485-9666

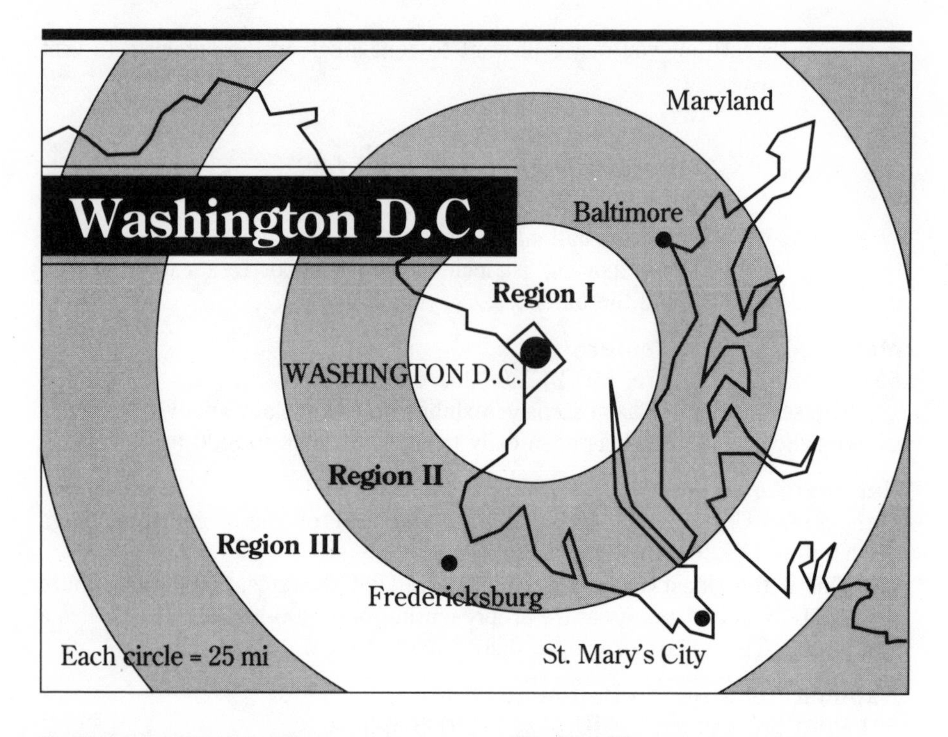

Indoor Activities

Region I

National Museum of Natural History of the Smithsonian Institute
National Aquarium

Region II

Maryland Science Center
National Aquarium in Baltimore

Outdoor Activities

Region I

Anacostia Park
Chesapeake and Ohio Canal National Historic Park
Constitution Gardens
Jug Bay Natural Area
Kenilworth Aquatic Gardens
Lady Bird Johnson Park—Lyndon Baines Johnson Memorial Grove
Merkle Wildlife Management Area
Rock Creek Park
Theodore Roosevelt Island
U.S. Botanic Garden
U.S. National Arboretum
Woodend Nature Trail

Region II

Baltimore Sewage-Treatment Plant
Battle Creek Cypress Sanctuary
Billy Goat Trail
Blockhouse Point
Bull Run Regional Park
Cedarville Natural Resources Management Area
Dranesville District Park
Great Falls Park
Mason Neck National Wildlife Refuge
McKee-Beshers Wildlife Management Area
Patuxent River State Park
Pohick Bay Regional Park
Prince William Forest Park
Riverbend Park
Sandy Point State Park
Seneca Creek State Park
Upper Rock Creek Regional Park
Wildcat Mountain Natural Area

Region III

Catoctin Mountain Park
Chancellor's Point Natural History Area
Cunningham Falls State Park
Westmoreland State Park

Bordering both sides of Anacostia River in the northeast and southeast sections of Washington, DC, this 750-acre park includes a bird sanctuary among its many features. It is open daily.

Baltimore Sewage-Treatment Plant
Route 150, Baltimore, MD 21224

Take the Beltway, Interstate 695, and get off at Eastern Avenue. Most people don't think of sewage treatment plants as natural areas, but excellent birding can be found here during the fall migration. Sunday is the best day to visit since no garbage trucks are in the way to disturb the birds. A tidal creek is nearby, and rain pools and landfill areas offer habitat to waterfowl and shorebirds.

Battle Creek Cypress Sanctuary
County Courthouse, Prince Frederick, MD 20678—301-535-1600

Take State Route 2/4 southeast from Washington to State Route 506 and Gray's Road. The 75- to 500-year old cypress trees in this sanctuary are some of the northernmost in the nation. A nature center contains natural history exhibits, and a quarter-mile boardwalk takes you through part of the swamp. From April through September, the sanctuary is open Tuesday–Saturday, 10:00 A.M.–5:00 P.M. and on Sunday, 1:00 P.M.–5:00 P.M. During the rest of the year, the sanctuary is open Tuesday–Saturday, 10:00 A.M.–4:30 P.M. and on Sunday, 1:00 P.M.–4:30 P.M.

Billy Goat Trail
c/o Park Superintendent, P.O. Box 158, Sharpsburg, MD 21782—301-739-4200

Located about one-half mile downstream from Great Fall's Tavern., this rugged trail circles Bear Island, the largest island below Great Falls. The falls can not be viewed from the trail, but their roar can be heard as you explore Mather Gorge, and the geological features of the river area. The trail is well marked, crosses several areas of exposed bedrock, and covers a five-mile circuit through forests and along the river's edge. Allow at least a half day to cover the complete trail, and wear sturdy shoes. Available at the visitors' center is *The River and The Rocks* by John Reed, Jr., a book that tells about the geology of the Billy Goat Trail and Great Falls' area.

Blockhouse Point
c/o Park Superintendent, P.O. Box 158, Sharpsburg, MD 21782—301-739-4200

Take Interstate 495 to Exit 39 and go west on River Road for just over ten miles to Pettit Road. A dirt-road trail begins to the left of the intersection. Huge rock outcroppings reach 100 feet above the canal and towpath, and great views of the Potomac River are provided from the top. The terrain around the outcroppings is rolling and heavily wooded. Many small birds can be found in the woods here, and waterfowl can be viewed along the river from above. The Dierssen Waterfowl Sanctuary is located about a mile downstream at Milepost 20.

Bull Run Regional Park
c/o Northern Virginia Regional Park Authority, 11001 Popes Head Rd., Fairfax, VA 22030—703-631-0550

Take Interstate 66 west from Washington to the Centerville Exit. Follow the signs to the park in Fairfax County, Virginia.

The spring wildflower display here may be one of the best in the region. Virginia bluebells blanket the fields and meadows in early spring. Masses of spring beauties, a small, delicate, white or pink flower, add a contrast to the bluebells. The one-and-a-half-mile, loop-shaped Bluebell Trail is best seen in mid-April, when other spring flowers of the region are also in bloom. An annual bluebell walk is held in April. Contact the Northern Virginia Regional Park Authority (see listing) for details. This is part of a chain of parks that have some heavily used sections, but much of the area is quiet and undisturbed, with over 40 miles of hiking trails. The park is closed from December 1st to March 15th.

Catoctin Mountain Park

Thurmont, MD 21788—301-824-2574

This 5,800-acre park is administered by the National Park Service, and adjoins Cunningham Falls State Park. It is in mountainous terrain, and reaches 1,621 feet in elevation. Over 90 percent of the land in both parks is forested with mixed deciduous trees. There is abundant wildlife. Nature talks and walks are given at the visitor center, and there are over 25 miles of trails in the park.

Cedarville Natural Resources Management Area

Route 4, P.O. Box 133, Brandywine, MD 20613—301-888-16212

Take Route 301 south from Washington. The area is four miles east on Cedarville Road and four miles north of Waldorf, Maryland.

This 3,689-acre refuge is on relatively flat land that offers varied habitat to wildlife. The headwaters of Maryland's largest freshwater swamp, Zekaih Swamp, is in the refuge. There is also a small bog with carnivorous plants. There are 15 miles of trails in the refuge. Campfire programs are held on Friday and Saturday nights at 8:00 P.M. from May to August.

Chancellor's Point Natural History Area

St. Mary's City, MD 20686

This area is about one hour southeast of Washington on the St. Mary's River and Chesapeake Bay. A loop trail leads you through 66 acres of habitat very similar to the one found here by settlers over 350 years ago. Woodland, marsh, beach, and bluff habitats are all included in this park. It is open daily.

Chesapeake and Ohio Canal National Historic Park

P.O. Box 158, Sharpsburg, MD 21782—301-739-4200

The park follows the Maryland shore of the Potomac River from Georgetown to Cumberland, Maryland. This 185-mile long, 1,000-foot wide park includes the canal that operated between 1828 and 1924, the towpath that ran beside it, and a number of parks, forests, and wildlife management areas. Bird watching is excellent year-round along most of the canal, and the lower end is especially popular. Although the 20 miles closest to Washington are heavily used, particularly on weekends, other sections have almost no traffic. There are several parks and trails along the canal that have special interest, including Billy Goat Trail and Blockhouse Point. See entries in this Outdoor Activities section.

Constitution Gardens
c/o National Park Service, 1100 Ohio Dr., SW, Washington, DC 20242—202-485-9666

These mall gardens cover almost 50 acres of lawns and hills. They are on Constitution Avenue between 17th and 23rd Streets, Northwest. The six-acre lake includes a one-acre island that has a footbridge leading to it. There are over 5,000 trees and 100,000 other plants in the garden, and bird life is abundant. The trails are easy and the gardens are open daily.

Cunningham Falls State Park
Thurmont, MD 21788—301-271-2495

Located three miles south of Thurmont on U.S. Route 15, this 4,500-acre park has rugged slopes, scenic forests, lakes, and waterfalls. See Catoctin Mountain Park, which adjoins Cunningham, for more information.

Dranesville District Park
c/o Fairfax County Park Authority, 4030 Hummer Rd., Annandale, VA 22003—703-941-5008

Take Interstate 495 to Exit 13, and follow Route 193, the Georgetown Pike, west to the parking lot. Over 175 species of wildflowers bloom in this park between mid-March and mid-June. They are at their peak from mid-April to mid-June, and continue through summer and fall in open woods and meadows. Five basic habitats can be found within the park, which is dominated by a mature oak and hickory forest. Over three miles of trails traverse the varied topography of the 385-acre park.

Great Falls Park
9200 Old Dominion Dr., Great Falls, VA 22066—703-759-2915

Take Interstate 495 to Exit 13. Follow Route 193—or the Georgetown Pike west to Route 738, Old Dominion Road. The visitor's center parking lot is about four miles farther north on Route 738.

A deep gorge and turbulent waterfalls characterize the fall line of the Potomac as it drops over 75 feet in a quarter of a mile at this transition from the Piedmont to the Coastal Plain. More than six miles of trails lead visitors through a variety of habitats that are home to large bird populations and wildflowers normally not found east of the mountains. The visitor center has many exhibits, including some on the geological activities that helped form the fall line. Guided walks are held year-round. The park is heavily used during summer months. The Fairfax County Park Authority has a marina and nature center upstream from Great Falls that can be reached by trail.

Jug Bay Natural Area
Croom Airport Rd., R.R. Box 3380, Upper Malboro, MD 20772—301-627-6074

An extensive system of trails and one-half mile of boardwalk lead through woodlands, swamp, and marsh in this wildlife management area. It offers one of the best bird-watching sites in the Chesapeake Bay region. Large numbers of birds of varied species are attracted here year-round. Of the 374 birds on the Maryland state list, over 80 have been seen nesting here. Nearly 250 total species have been recorded as seen in the area. The appearance of the area changes dramatically with the seasons.

Since this area along the Patuxent River is managed primarily to protect the river and wildlife, all visitors must register in advance for orientation to the area. Flat-water canoeing is excellent in the area, and canoes can be rented from the park by reservation. The park office is open from 8:00 A.M. to 4:30 P.M., Tuesday through Saturday.

Kenilworth Aquatic Gardens
c/o National Park Service, 1100 Ohio Dr., SW, Washington, DC 20242—202-485-9666

Located at New York and Kenilworth Avenues, Northeast, these gardens contain 44 ponds filled with over 100 varieties of water plants. Birds, insects, and amphibians may be found here. Riverside life also can be seen. The gardens are open daily from 7:00 A.M. to 5:00 P.M.

Lady Bird Johnson Park—Lyndon Baines Johnson Memorial Grove
c/o National Park Service, 1100 Ohio Dr., SW, Washington, DC 20242—202-485-9666

This park is located on George Washington Memorial Parkway on what was previously known as Columbia Island at the Virginia end of the Arlington Memorial Bridge. Dedicated in 1968, it now is noted for the 15-acre grove of white pine, dogwoods, azaleas, and rhododendron that form a natural backdrop for the LBJ Memorial. In the spring as many as one million daffodils bloom here, and are followed by 2,700 dogwoods. Birds are abundant most of year. The park is open daily.

Mason Neck National Wildlife Refuge
9502 Richmond Highway, Suite A, Lorton, VA 22079—703-339-5278

Located south of Washington, take Interstate 95 for 25 miles to the Lorton Exit. Follow State Route 242 or Gunston Road to High Point Road.

This is the only national wildlife refuge established especially for bald eagles. This 1,117-acre refuge is located on a peninsula in the Potomac River. It includes 600 acres of upland forest and 300 acres of riverside marsh. Part of refuge is closed to protect eagles from disturbances, and the only public trail is the three-mile Woodmarsh Trail that skirts the Great Marsh. This is an interpretive trail. The entire refuge is closed from December 1st to March 31st. Guided walks and canoe tours are sometimes arranged in advance.

McKee-Beshers Wildlife Management Area
c/o District Manager, Strider Work Center, 12510 Clopper Rd., Germantown, MD 20767—301-428-0438

Take Interstate 495 to Exit 39. Go west on River Road through Potomac. After about 15 miles, watch for Hunting Quarter Road on the left. This area is next to Seneca Creek State Park.

Known for its excellent bird-watching opportunities, McKee-Beshers Wildlife Management Area is a 1,500-acre refuge adjoined to the Chesapeake and Ohio Canal. It attracts ducks, herons, hawks, and warblers, among a wealth of other birds. Woodland and field wildflowers burst forth from early spring to mid-summer. You may slog through the swamps, cross the fields, and meander through the woods in search of wildlife. Summers here are very hot and humid, and mosquitoes and ticks can be hazardous.

Merkle Wildlife Management Area

c/o Cheltenham Work Center, 11000 Old Indian Head Rd., Upper Malboro, MD 20870—301-372-8128

Take U.S. Route 301 south from Upper Malboro to State Route 382. Then go left to Croom Airport Road.

This 1,223-acre refuge is close to downtown Washington, and is one of the last undeveloped areas along the Patuxent River in Prince Georges County. Noted for its concentration of wintering waterfowl, with 5,000–10,000 Canadian geese normally in residence, the refuge has almost 500 acres of tidal marsh as well as forested uplands. Bald eagles nest along the river, and captive-hatched peregrine falcons have been released here. Hiking is good on the seven miles of gravel roads in the refuge.

Patuxent River State Park

17400 Annapolis Rock Rd., Woodbine, MD 21797—301-489-4656

Near the Triadelphia Reservoir, this 12-mile long, 6,000-acre park is largely undeveloped shoreline. There are no marked trails in the park, but many visitors enjoy hiking along the river. Wildflowers, forests, and bird life are all abundant.

Pohick Bay Regional Park

c/o Northern Virginia Regional Park Authority, 11001 Popes Head Rd., Fairfax, VA 22030—703-339-6100

Located south of Washington, follow Interstate 95 for 25 miles to the Lorton Exit. Then follow the signs to Gunston Hall.

This 1,000-acre park is located next to Mason Neck National Wildlife Refuge on a peninsula in the Potomac River. It has 12 miles of hiking trails, including five miles along Pohick Creek outside the park. The area is heavily used in season.

Prince William Forest Park

P.O. Box 208, Triangle, VA 22172—703-221-7181

Located south of Washington, drive 32 miles on Interstate 95. Then exit on State Route 619.

This 17,348-acre park adjoins the U.S. Marine Corps Reservation at Quantico Station, Virginia. Its picnic and camping facilities are often crowded, but its 35 miles of hiking trails and roads lead to uncrowded places. There you can enjoy the mixed forests that are home to wild turkey, ruffed grouse, many songbirds, and mammals such as deer, foxes, beavers, and flying squirrels. There are nature trails, a nature center, guided hikes, and illustrated talks daily in the summer. These are available on weekends during the rest of the year.

Riverbend Park

8814 Jeffery Rd., Great Falls, VA 22066—703-759-3211

Take Interstate 495 to Exit 13. Go west on Georgetown Pike to Riverbend Road. Turn right on Jeffery Road.

This park is located on a floodplain of the Potomac River above Great Falls. In the spring wildflowers carpet the ground, and migrating songbirds serenade you from the cover of new, green foliage. Wood ducks, cuckoos, and hummingbirds, among others, nest here in the summer. Fall brings bright colors and the park becomes an excellent foliage walk.

Although presenting some of the same habitats, Riverbend is less crowded than nearby Great Falls Park. A one-and-a-half-mile trail connects the visitor centers of the two parks. Riverbend has trails, naturalist-directed activities, and a nature center. It is open year-round.

Rock Creek Park and Nature Center

5200 Glover Rd., NW, Washington, DC 20015—202-426-6829

The park is four miles long and one mile wide, and it runs almost the entire length of the city. It includes large stands of hardwoods, with wooded valleys and bottomland within its 1,800 acres. Human activities have greatly changed the natural features of the park. While there are many exotics found throughout the park, the plant communities have been least affected by the surrounding urban development. The streams are so polluted that few fish live in them, and many native mammals no longer live there. The bird population is varied, but the park is far from the best birding site in the city.

The northern end of the park is less used, and visitors have a better chance of enjoying nature there. The 15 miles of nature trails give visitors opportunities to leave some of the more heavily used parts of the park. The Rock Creek Nature Center features wildlife displays, exhibits, nature walks, and an observation deck.

Sandy Point State Park

800 Revell Highway, Annapolis, MD 21401—301-757-1841

Take State Route 50 east from Washington to the last exit before the Chesapeake Bay Bridge. Then follow the signs to the park.

Nearly every bird found in Maryland has been seen at this park. Two factors make this one of the best birding spots on Maryland's western shore. The first is that the point on which the park sits juts out into a narrow section of the Chesapeake Bay, and migrating birds are attracted to the point's prominence. The second is the wide variety of habitat in the 679-acre park. In addition to beach and open water, the park has a pond that was dredged for boating, mud flats, scattered woodlands, and both fresh and brackish marshes.

Waterfowl flock to the bay and pond in the fall, and many remain there over the winter. Shorebirds are numerous in the spring and fall. An observation platform on one of the dikes offers a good view of the mud flats and bay.

Hawks pass through the area in the fall, and on some days as many as 4,000 fly overhead. Peregrine falcons have wintered on the Chesapeake Bay Bridge for the past decade, and some are reported to be nesting there.

This park is open year-round, and the best time to visit it is after the summer crowds have left on Labor Day, and before they return on Memorial Day. During this period you can see the most birds, and the fewest people.

Seneca Creek State Park

P.O. Box 2235, Gaithersburg, MD 20760—301-924-2127

Located next to McKee-Beshers Wildlife Management Area, this park is another excellent bird-watching spot. It offers a variety of habitat.

Theodore Roosevelt Island

Turkey Run Area, George Washington Memorial Pkwy., McLean, VA 22101—202-426-6922

Located on Potomac River just north of the Theodore Roosevelt Bridge, this island is accessible only from the northbound lanes of the George Washington Memorial Parkway.

More than 50 species of trees, 200 varieties of wildflowers, and many types of birds and mammals can be found in the marsh, swamp, and hardwood forest on this 88-acre island. This three-quarter-mile long and one-quarter-mile wide island has at least eight different habitats, and nearly three miles of trails lead visitors through most of them. The swamp areas are almost impassable after heavy rains, and appropriate footwear should be worn. Guided nature hikes are given on weekends.

U.S. Botanic Garden

c/o National Park Service, 1100 Ohio Dr., SW, Washington, DC 20242—202-225-8333

At First Street and Maryland Avenue, Southwest, this garden can be visited even in inclement weather as the conservatory contains many rare, native plants. There is also an outdoor garden park. It is open daily from 9:00 A.M. to 5:00 P.M. During the summer the garden is open 9:00 A.M.–9:00 P.M.

U.S. National Arboretum

3501 New York Ave., NE, Washington, DC 20002—202-472-9100

This 444-acre arboretum contains many plants commonly grown in the eastern United States. A network of trails and roads lead through thousands of woody plants. Birds can be seen here in season. The arboretum is open Monday–Friday, 8:00 A.M.–5:00 P.M., and Saturday–Sunday, 10:00 A.M.–5:00 P.M.

Upper Rock Creek Regional Park

6700 Needwood Rd., Rockville, MD 20855—301-948-5053

Located about one-half hour northwest of Washington, over four miles of trails crisscross this 2,000-acre park. The Meadowside Nature Center has many natural history exhibits. Lake, creek, and meadow habitats can be found here. The park is open daily from 8:00 A.M. to sunset. The center is open Tuesday–Saturday, 9:00 A.M.–5:00 P.M., and Sunday, 1:00 A.M.–5:00 P.M.

Westmoreland State Park

Route 1, P.O. Box 53-H, Montross, VA 22520—804-564-9057

Go south on Interstate 95 from Washington to Fredericksburg. Follow State Route 3 east to the park.

This 1,302-acre park is located on the Potomac River between the birthplaces of George Washington and Robert E. Lee. Sandy beaches and the cliffs above have many fossils which can not be removed. A large portion of the park is reserved as a natural area. The park is heavily used for water sports in summer, but the 17 miles of trails are often unused. The Big Meadow Nature Trail includes an observation tower in Yellow Swamp.

Wildcat Mountain Natural Area

c/o The Nature Conservancy, 1800 N. Kent St., Arlington, VA 22209—703-524-3151

Take Interstate 66 about 35 miles past the Beltway. Exit on U.S. Route 17, then turn right on State Route 691 and go about five miles.

This natural area is on the western slope of the Wildcat Mountain, in the foothills of the Blue Ridge Mountains. The 650 acres include vegetation typical of the southern Piedmont. Different stages of natural succession are on display as the area is being reclaimed by the forests of oak and hickory. More than 170 types of wildflowers and 95 species of birds can be found here. Greenstone, a low-grade metamorphic rock, can be found throughout the park. Deer, raccoons, and foxes can be found in the clearing around the small human-made pond that was abandoned, and is slowly being encroached upon by the surrounding forest.

Woodend Nature Trail

8940 Jones Mill Rd., Chevy Chase, MD 20815—301-652-9188

This trail, located at the headquarters of the Audubon Naturalist Society of the Central Atlantic States, is part of a 40-acre estate owned by the Society. The estate is a wildlife haven in the middle of a suburb, and counts numerous small mammals and birds among its residents.

Organizations That Lead Outings

Audubon Naturalist Society
8940 Jones Mill Rd., Chevy Chase, MD 20815—301-652-9188

Potomac Appalachian Trail Club
1718 N St., NW, Washington, DC 20024—202-638-5306

The Sierra Club
1863 Kalorama NW, Apt. 1B, Washington, DC 20009—202-547-5551

A Special Outing

It is a narrow strip of land, averging less than 1,000 feet across, that extends from Georgetown about 185 miles to Cumberland, Maryland. The Chesapeake and Ohio Canal National Historic Park offers visitors to the Washington area an excellent opportunity to explore the Maryland shore of the Potomac River.

The canal, which was started in 1828 and finished in 1850, operated from 1850 to 1924. Then the canal was damaged so severely by a flood that it could no longer be used. Today the canal and its towpath are a national historic park, and along with adjacent parks, forests, and wildlife management areas, provide many naturalist opportunities to visitors.

Along the lower end of the canal bird-watching is excellent, especially in the spring and fall when warblers, vireos, thrushes, and hawks, among others, use the canal as a rest stop on their long migrations. During the winter many waterfowl use the river, and the deciduous forests bordering the canal are nesting grounds for many resident birds during late spring and summer.

Although the canal route was engineered, much of it offers a wilderness experience for any who hike its extensive sections. While the lower end of the canal close to Washington is heavily used, particularly on weekends, it is less crowded the farther away you get from the city.

The terrain near Washington is rolling fields, but these soon give way to more mountainous terrain. At one point the canal and towpath go through a mountain via the half-mile long Paw Paw Tunnel.

Black bears, bobcats, beavers, muskrats, red foxes, raccoons, copperheads, and snapping turtles all can be found along the path and among the mixed deciduous forests that border the canal and river. Spring brings wildflower blooms such as violets, Virginia bluebell, jack-in-the-pulpit, and wild geranium.

Visitor centers are located at a number of stops along the canal and evening programs and guided walks are offered all year. Campsites are spaced about every five miles, and many visitors hike or bike the entire distance. Others find one of the many entry points and enjoy short excursions to sites such as Widewater, Great Falls, Whites Ferry, and Paw Paw Tunnel. No road parallels the canal, but a number of spur roads lead to it from U.S. Route 40 to entry points.

For more information about the canal contact: The American Youth Hostels, 1520 16th Street, Northwest, Washington, DC 20036; 202-783-4943; the Superintendent, Chesapeake and Ohio Canal National Historic Park, P.O. Box 158, Sharpsburg, MD 21782; 301-739-4200; or the Great Falls Tavern, 11710 MacArthur Boulevard, Potomac, MD 20854; 201-229-3613. The American Youth Hostel has maps and towpath guides of the entire canal.

Nature Information

Washington abuts both Virginia and Maryland. Agencies from those two states and their counties, as well as those of Washington and the National Park Service, share responsibility for many parks and natural sites in the region. Some of these follow.

Fairfax County Park Authority
4030 Hummer Rd., Annandale, VA 22003—703-941-5008

Maryland Park Service
Tawes State Office Bldg., Annapolis, MD 21401—301-269-3761
Available publications include *Maryland: The Mountains, the Bay, the Ocean* and *Exploring Nature and History in Maryland's State Parks.*

National Parks Service
U.S. Department of the Interior, 1100 Ohio Dr., SW, Washington, DC 20242—202-485-9666

Northern Virginia Regional Park Authority
11001 Popes Head Rd., Fairfax, VA 22030—703-278-8880

Washington, DC Convention and Visitors Association
1212 New York Ave., NW, 6th Floor, Washington, DC 20001—202-789-7000
(Call or write only; no walk-in requests.)

Washington Information Center
1445 Pennsylvania Ave., NW, Washington, DC 20004—202-789-7000

Further Reading

Fisher, Alan. *Country Walks Near Washington.* Boston: Appalachian Mountain Books, 1981.

Meanley, Brooke. *Marshes of the Chesapeake Bay Country.* Centreville, MD: Tidewater Publishers, 1975.

———. *Waterfowl of the Chesapeake Bay Country.* Centreville, MD: Tidewater Publishers, 1982.

Reed, John C., Jr., Robert S. Sigafoos, and George W. Fisher. *The River and the Rocks: The Geologic Story of Great Falls and the Potomac River Gorge.* U.S. Geological Survey, Bulletin No. 1471. Washington, DC: Government Printing Office, 1980.

Robbins, Chandler S., and Danny Bystrak. *Field List of the Birds of Maryland.* 2nd ed. Maryland Avifauna No. 2. Baltimore: Maryland Ornithological Society, 1977.

Schubel, John R. *The Living Chesapeake.* Baltimore: Johns Hopkins Press, 1981.

Shosteck, Robert. *Potomac Trail and Cycling Guide.* Oakton, VA: Appalachian Books, 1976.

Silderhorn, G.M. *Tidal Wetland Plants of Virginia.* Gloucester Point, VA: Virginia Institute of Marine Science, 1976.

Thomas, Bill, and Phyllis. *Natural Washington.* New York: Holt, Rinehart & Winston, 1980.

White, Mel. *A Guide to Virginia's Wildlife Management Areas.* Richmond: Virginia Commission of Game and Inland Fisheries.

Wilds, Claudia. *Finding Birds in the National Capital Area.* Washington, DC: Smithsonian Institution Press, 1983.

SOUTHEAST

Atlanta
Tampa–St. Petersburg
Memphis
New Orleans

There is more geographical and natural diversity among the metropolitan areas of the Southeast than in the Northeast. Of the four southeastern cities represented in this guide: one is located on the Piedmont; one on the Gulf Coast of Florida; one on the banks of the Mississippi River; and the last on the delta flood plains of the lower Mississippi. The cultural and social diversity of these four is as great as their geographical locations.

Atlanta is representative of the New South, and offers visitors much the same buoyant ambience as other growing Sunbelt cities. New Orleans, on the opposite end, is still very much part of the Old South, with an ambience in tune with the humidity and hot weather that frequently slows visitors and residents to a snail's pace.

The choice for outdoor recreation in the region has long been hunting and fishing. For many years most southern states made little effort to set aside areas as wildlife refuges and sanctuaries for naturalists who wished only to look at and enjoy nature. Florida was a striking exception, as it began early to set aside prime areas of wilderness.

The rapid social and economic changes in the region during the past quarter century have been accompanied by changes in attitudes toward the environment. More states have begun to protect the Southeast's remaining natural sites, and to open them to visitors and residents who want to enjoy nature without destroying fragile habitats.

Natural History of the Southeast

As noted earlier, the Southeast has a tremendously diverse geography. Atlanta sits on a rolling plain of the Piedmont at the base of the Appalachian Mountains. It has many nearby forests with the widest variety of trees found in the country. A major river flows through it. Atlanta is blessed with more moderate summer temperatures than any of the other three cities in this section.

Tampa sits on the west coast of Florida, and is influenced greatly by the Gulf of Mexico. It offers visiting naturalists an exceptional opportunity to enjoy the seacoast and its many activities, as well as observe the environment offered by slow-moving rivers and swamps farther inland.

Memphis was built on the bluffs of the Mississippi River, and sits on the flatlands of western Tennessee. Its naturalist opportunities center around the riparian environment of a major river, and of the pine and hardwood forests that ring low-lying swamps in Tennessee, and nearby Arkansas and Mississippi.

New Orleans has water—from the Mississippi River, from the swamps of the delta that extends across southern Louisiana, and from Lake Pontchartrain, which lies east of the city. All of these combine to give New Orleans more swamps inside its city limits than most states have. They also limit naturalist activities almost exclusively to a single environment.

The heat and humidity common to the Southeast is found in all four cities. Therefore, naturalist activities in the region are best enjoyed during the long fall and spring seasons. Most of the time, pleasant winters can be enjoyed outdoors around Tampa and in New Orleans. Sometimes, however, it is wet, windy, and chilly during the winter. While Atlanta and Memphis do not have the long bone-chilling winters of the Northeast and Midwest, they do have enough cold to make outdoor activities unpleasant during some periods.

To better understand the naturalist activities around each southeastern city, you should read about: hardwood forests and riparian life zones near Atlanta; seashore, salt marsh, and freshwater swamps near Tampa; lowland hardwood and pine forests, swamps, and riparian life zones around Memphis; and swamps in New Orleans.

Less specific material has been written about these cities, and the Southeast in general, than about the Northeast. Therefore, more information has to be gleaned from general naturalist books to learn about the flora and fauna of the region.

Further Reading

Green, Wilhelmina F., and Hugo Blumquist. *Flowers of the South.* Chapel Hill, NC: University of North Carolina Press, 1953.

Atlanta

Atlanta is the largest center of commerce in the South. While maintaining a reputation as a pleasant metropolitan area where industry and commerce have brought dozens of shining skyscrapers, along with a healthy economy, it has found itself confronted with many of the problems shared by most large industrial and commercial centers. Commuters clog both the expressways and downtown

ABOVE: Visitors to the Fernbank Science Center can hike to shaded ponds in the midst of mixed hardwood forests. Photo by Bernard J. Thoeny.

streets; urban sprawl is pushing ever outward, and contact with nature is becoming increasingly difficult for residents and visitors.

Some respite from these pressures, however, is given by the natural setting of the city. It is blessed with stands of hardwood forests, and its suburbs are noted for their wooded streets and spring blooms of dogwoods and azaleas. These, along with a moderate climate, help make Atlanta a pleasant city to visit, even though nature activities close by are more limited than in other American cities.

Its geographical setting also adds much to outdoor enjoyment. Atlanta sits on the rolling Piedmont that slopes gently from the base of the Appalachian uplift. The origin of many of Georgia's major rivers is nearby. The fall line of these lies to the southeast of Atlanta where the Piedmont gives way to the coastal plain.

These geographical characteristics account for the variety of naturalists' outings that are available in and around Atlanta. Swiftly flowing streams and rivers can be found within reasonable distances from the city. The Chattahoochee River National Recreation Area actually extends into the city limits. Hardwood forests with an abundance of bird and mammal life are close by. The geological wonders of Stone Mountain and similar granite outcroppings give visitors opportunities to study the geological history of the southern Appalachian Mountains.

Climate and Weather Information

Atlanta's elevation of approximately 1,000 feet helps moderate the hot, humid summers that oppress much of the Southeast. It brings somewhat colder winters to Atlanta than other southeastern cities experience. Both fall and spring are long seasons, and excellent times to be outdoors. Be prepared for rain between July and August and between February and March.

Getting Around Atlanta

As in other major cities, rush hour is to be avoided in Atlanta. The main streets downtown are heavily congested during this time, and the expressways are jammed with commuter traffic. Construction projects are ongoing problems along Atlanta's expressways as they are constantly being widened. However, a rapid transit system is being developed.

Since it is not laid out in the traditional grid system, the city has few rectangular blocks. Overuse of the word *Peachtree* in street names can cause visitors some navigational problems. On-street parking is minimal downtown, but off-street parking is plentiful, if somewhat expensive. Taxis are plentiful, with reasonable rates, and rental car companies have offices both at the airport and downtown.

Public transportation is adequate, and an attempt is being made to improve it as the Metro Atlanta Rapid Transit Authority is expanding routes and constructing an extensive monorail system. Only a few of the outdoor activities listed can be reached by public transportation.

Indoor Activities

In addition to the following organizations, colleges and universities in and near Atlanta have museums that frequently feature natural science exhibits. These include Emory, Georgia Tech, and Atlanta Universities. Others are often listed in local newspapers.

Atlanta Botanical Gardens
South Prado Loop Rd., Atlanta, GA 30303—404-876-5858

In addition to outdoor gardens, there is a tropical plant conservatory here, plus a library and exhibition hall. The gardens are open Monday–Saturday, 9:00 A.M.–dusk and Sunday, noon–dusk.

Chattahoochee Nature Center
9133 Willeo Rd., Roswell, GA 30075—404-992-2055

See entry in Outdoor Activities for more information. The center is open Monday–Saturday, 9:00 A.M.–5:00 P.M. and Sunday, 10:00 A.M.–5:00 P.M.

Fernbank Science Center
156 Heaton Park Dr., NE, Decatur, GA 30307—404-378-4311

The center includes a planetarium, natural history museum, and green houses. It is open Tuesday–Friday, 8:30 A.M.–10:00 P.M., Monday, 8:30 A.M.–5:00 P.M., Saturday, 10:00 A.M.–5:00 P.M., and Sunday, 1:00 P.M.–5:00 P.M. See entry in Outdoor Activities for more information.

State Botanical Gardens of Georgia
University of Georgia, 2450 S. Milledge Ave., Athens, GA 30605—404-542-6329

See entry in Outdoor Activities for more information. The gardens are open Monday–Saturday, 9:00 A.M.–4:30 P.M., and Sunday, 11:30 A.M.–4:30 P.M.

Outdoor Activities

Although Atlanta offers its visitors many state parks, recreational activities, and historical sites nearby, few of these have extensive nature activities. Some of the state parks do offer hiking trails that lead visitors away from the crowds.

Allatoona Lake
P.O. Box 487, Cartersville, GA 30120—404-382-4700

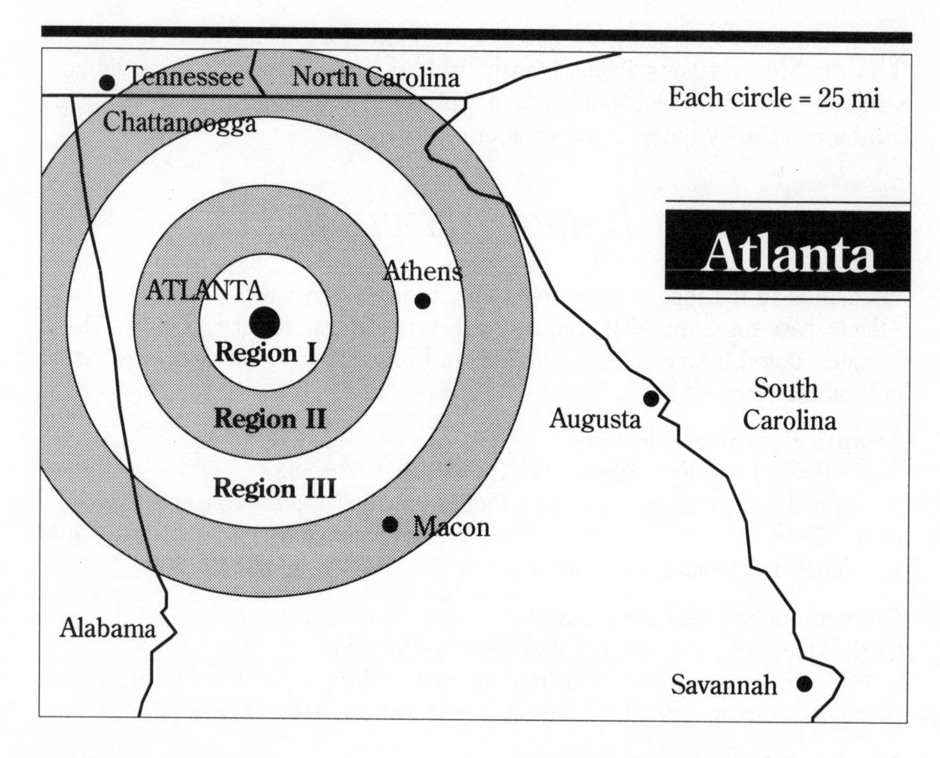

Atlanta

Indoor Activities

Region I

Atlanta Botanical Gardens
Chattahoochee Nature Center
Fernbank Science Center

Region III

State Botanical Gardens of Georgia

Outdoor Activities

Region I

Atlanta Botanical Garden
Chattahoochee Nature Center
Chattahoochee River National Recreation Area
Fernbank Science Center
Panola Mountain State Conservation Park
Stone Mountain Park
Yellow River Wildlife Game Ranch

Region II

Allatoona Lake
Sweetwater Creek State Park

Region III

Amicalola Falls State Park
Chattahoochee National Forest
State Botanical Gardens of Georgia

Although this dam was built for flood control and power generation, the area has heavy recreational use. There are numerous old logging roads that traverse the upland and bottomland hardwood and pine forests where many birds and mammals can be found. Otters, raccoons, bobcats, deer, herons, egrets, hawks, and occasionally bald eagles can all be seen in the area. The visitor center has camp-fire programs and a fall leaf tour.

Amicalola Falls State Park
Star Route, Dawsonville, GA 30534—404-265-2885

This 296-acre park sits high in the Blue Ridge Mountains near Dahlonega. It includes the highest waterfall in Georgia. Amicalola Falls is a 729-foot series of falls that can be viewed from an overlook platform. The park is densely forested, and has trails to the falls and around a small lake. A trail leads from the falls to the beginning of the Appalachian Trail on Springer Mountain seven miles away. At an elevation of 2,600 feet, the park offers a respite from hot summer days.

Atlanta Botanical Garden
South Prado Loop Rd., Atlanta, GA 30303—404-876-5858

The garden, which is located in Piedmont Park, includes 53 acres. Exhibits and vegetable, herb, Japanese, and rose gardens are featured. Gardening classes and workshops are offered frequently.

Chattahoochee National Forest
P.O. Box 1437, Gainesville, GA 30501—404-536-0541

This national forest is the southern extension of the most extensive hardwood forest in the world. Several points within the forest offer excellent natural outings. The first is Brasstown Bald. At 4,784 feet, it is the highest point in Georgia. A visitor center on top of the bald includes exhibits, a theater, and an observation deck. There are also three hiking trails near the summit.

The second point of interest is Anna Ruby Falls Scenic Area, where two creeks converge as they enter Unicoi Lake. The two falls are the site of several hiking trails and a picnic area. There is a state park at Unicoi Lake.

The last point of interest is the Cahutta Wilderness Area. It offers extensive hiking opportunities within its 32,000 acres.

Chattahoochee Nature Center
9133 Willeo Rd., Roswell, GA 30075—404-992-2055

There is an abundance of nature trails, with native plants and wildlife in the Chattahoochee National Recreation Area. The center of nature activities is the Chattahoochee Nature Center. This private, nonprofit organization offers wildlife and plant exhibits, special programs, nature walks, and lectures. It is housed in a 3,500-square-foot lodge that has a garden of native plants leading to its entrance, and a 1,400-foot raised boardwalk leading from it through a marsh toward the river.

Chattahoochee River National Recreation Area

P.O. Box 1396, Smyrna, GA 30080—404-952-6009

While most rivers that run through or near large metropolitan areas are befouled beyond use, the Chattahoochee is an exception. It borders Atlanta, and is the principal waterway around the city. The residents have long enjoyed the many activities the river provides, and have protected it.

In 1978, Congress created the National Recreation Area that extends along 48 miles of the river. This area is protected and preserved, to provide visitors a wide variety of activities, including hiking and bird watching. The Cochran Trail is the most popular within the area, but it is often overcrowded and noisy.

Fernbank Science Center

156 Heaton Park Dr., NE, Decatur, GA 30307—404-378-4311

In the midst of a northeastern Atlanta residential area, the 65-acre Fernbank Forest offers visitors a look at a virgin forest, and two miles of trails, including an easy effort trail for those with physical handicaps. Another trail provides opportunities for the visually impaired to experience native flora. The forest is part of the Fernbank Science Center, which includes an exhibit building, an observatory, and a planetarium.

Panola Mountain State Conservation Park

Route 1, P.O. Box 364-B, Stockbridge, GA 30281—404-474-2914

Located just 18 miles southeast of Atlanta on State Route 155, this park is small, only 537 acres. But it has been carefully managed to preserve the area. The mountain is a granite monadnock. It is surrounded with a pine-hardwood forest, old fields, and a three-acre pond fed by a mountain stream. The interpretive center offers guided walks in park areas that are closed to unaccompanied visitors. There are also self-guided nature trails.

State Botanical Gardens of Georgia

University of Georgia, 2450 S. Milledge Ave., Athens, GA 30605—404-542-6329

This 293-acre facility sits on the banks of the Middle Oconee River. It has nature trails with a wide variety of native plants, flowers, and trees, a visitor center, and a conservatory. The gardens are open Monday–Saturday, 9:00 A.M.–4:30 P.M. and Sunday, 11:30 A.M.–4:30 P.M.

Stone Mountain Park

P.O. Box 778, Stone Mountain, GA 30086—404-498-5600

While most of the area surrounding this 3,200-acre recreational and historic park is devoted to activities other than those of nature, there are several interesting areas. A one-and-one-third-mile trail leads to the summit of this massive granite dome that stands 825 feet above the surrounding plain. The dome is 300 million years old, measures five miles in circumference, and covers 583 acres. An eight-mile trail leads hikers around the dome. The Wildlife Trail leads visitors through a woodland setting with running streams and some native and domestic animals in a small zoo.

Sweetwater Creek State Park
Route 1, Mt. Vernon Rd., Lithia Springs, GA 30057—404-944-1700

This park is off Interstate 20, about 25 miles from downtown Atlanta, in one of the most rapidly developing counties surrounding the city. A 250-acre reservoir, more than five miles of nature trails, and a small rapids all make this an excellent place to visit. The pine and hardwood forests are home to many birds, rabbits, and squirrels.

Yellow River Wildlife Game Ranch
4525 U.S. Route 78, Lilburn, GA 30247—404-972-6643

Native wildlife of Georgia, such as deer, raccoons, and bobcats, can be observed in this 24-acre animal preserve. Ranch officials are on hand to answer questions and to help visitors gain an understanding of the delicate ecological balance necessary for the survival of native species. While this ranch is really a zoo, and has many of the normal zoo trappings, it does offer excellent insight into the way animals lived in the wild before humans invaded the dense forests of northern Georgia.

Organizations That Lead Outings

Most of these organizations lead outings some time during the year. It is best to check in your area of interest.

Audubon Society of Atlanta
2504 Brookdale Dr., NE, Atlanta, GA 30345—404-321-6079

Chattahoochee Nature Center
9133 Willeo Rd., Roswell, GA 30075—404-992-2055

Fernbank Science Center, 156 Heaton Park Dr., NE, Decatur, GA 30307—404-378-4311

National Wildlife Federation—Southeast
Natural Resources Center, 1718 Peachtree St., NW, Atlanta, GA 30309—404-876-8733

Nature Conservancy
4725 Peachtree Corners Circle, Atlanta, GA 30309—404-263-9225

Sierra Club
241 Marietta St., NW, Atlanta, GA 30318—404-584-8502

The Wilderness Society
1819 Peachtree Rd., NE, Atlanta GA 30309—404-355-1783

A Special Outing

The Appalachian Trail is the most famous trail in America, and its 2,000 plus miles pass through the most densely populated regions of America. Thousands of hikers walk some part of the trail each year, and most rave about the scenic beauty of the mountains and the closeness with nature they feel.

The Appalachian Trail does not pass near Atlanta. It does, however, end less than 100 miles northeast of the city. Visitors to Atlanta have opportunities to explore the Chattahoochee National Forest, which includes several wilderness areas, state parks, a designated Wild and Scenic River, beautiful waterfalls, and the beginning of (or the end of, depending on your orientation) the Appalachian Trail.

By driving a little over an hour, enjoying a rest at Amicalola Falls State Park, and following a blue-blazed trail for seven miles to Springer Mountain in northern Georgia, you can reach the southern terminus of the Appalachian Trail. The trail from Amicalola Falls goes through dense forests, across clear mountain streams, and along scenic ridges on its way to Springer Mountain. If you choose to spend only a day on this outing, you will not have much hiking time on the Appalachian Trail. But, if you make a fast hike, you should be able to at least have lunch on the trail before returning to Amicalola Falls.

For those interested in using this route to reach Springer Mountain, contact the headquarters at Amicalola Falls State Park for a trail map and information on hiking conditions. See the Outdoor Activities section for the address and telephone number.

Nature Information

Chattahoochee National Forest
P.O. Box 1437, Gainesville, GA 30501—404-536-0541

This national forest provides literature on various aspects of the region, including a list of trails with mileages. The *Field Guide to Wildlife on the Chattahoochee National Forest* is also available.

Department of Natural Resources
Communications Office, 205 Butler St., SE, Suite 1258, Atlanta, GA 30334
800-542-PARK outside Georgia; 800-342-PARK inside Georgia

Georgia Department of Industry and Trade
Tourist Division, P.O. Box 1776, Atlanta, GA 30301—404-656-3590

Parks and Historic Sites Division
Department of Natural Resources, 270 Washington St., SW, Atlanta, GA 30334—404-656-3530

Further Reading

Hans, Daniel W., ed. *A Birder's Guide to Georgia.* Atlanta: Georgia Ornithological Society, 1975.

Sullivan, Jerry, and Glenda Daniel. *Hiking Trails in the Southeast Mountains.* Chicago: Contemporary Books, 1975.

Tampa–St. Petersburg

Florida is noted for its many recreational opportunities, and it caters heavily to out-of-state tourists. Unlike many regions that attract tourists, however, Florida has not neglected its natural areas. While there has been heavy development throughout much of the state, Florida still has some of the wildest country in the East, and it is second only to New Hampshire among eastern states in the percentage of land devoted to wilderness.

Interest in preserving natural sites began early in Florida. The first national wildlife refuge was established on Pelican Island during Teddy Roosevelt's administration. The first national forest in the East—the Ocala—opened in 1903.

ABOVE: The swamps and hammocks that border Alice Creek as it empties into Hills Bay are typical of the West Florida coast. Photo by SWIM Department, Southwest Florida Water Management District.

These efforts continue today, and visitors to Florida have many opportunities to enjoy nature. There are more than 1,000 miles of developed trails in parks and forests in the state, plus the extensive Florida Trail, which one day will reach from one end of the state to the other. There are also 35 rivers and waterways throughout the state that have been designated as the Florida Canoe Trail System.

The vegetation and wildlife of these wilderness areas have characteristics of both temperate and tropical zones. The state is known as a bird-watcher's wonderland since more than 400 varieties of birds have been recorded there.

Many of the parks, refuges, trails, and sanctuaries can be reached from the Tampa–St. Petersburg area. While it takes about ten hours to drive the length of Florida, it is takes less than three hours to drive from the Gulf Coast near Tampa to the Atlantic Coast. This short distance puts most natural areas of central Florida within reasonable driving distance of Tampa.

Climate and Weather Information

Outdoor recreation is a year-round activity in Florida. Average high temperatures in Tampa are always in the 70s or higher, and lows seldom drop below 50°F even in the middle of the winter.

The only time inclement weather interferes with outdoor activities is during the peak of summer when the high temperature averages 90°F. When this heat is accompanied by the summer rains and thunderstorms that occur between June and September, few people want to engage in strenuous activities.

Getting Around Tampa–St. Petersburg

These cities are relatively easy for visitors to get around in. Their streets follow a grid system, and are laid out according to the compass. Streets and ways run north and south, and avenues, terraces, and places run east and west.

Rush hour can be difficult, especially if attempting to cross either of the three bridges that connect St. Petersburg with the mainland. Parking is limited on streets, and many lots are metered. Taxis with average rates are air conditioned and easy to find. Public transportation is adequate, but only a few of the activities listed are served by it.

Indoor Activities

There are some indoor activities in the Tampa Bay region that can be enjoyed any time of the year. They are few compared to the numerous outdoor ones.

Marie Selby Botanical Gardens
811 S. Palm Ave., Sarasota, FL 34236—813-366-5730

These gardens are on 16 acres of waterfront property, and are known for a collection of tropical plants, especially epiphytes or air plants. The Tropical

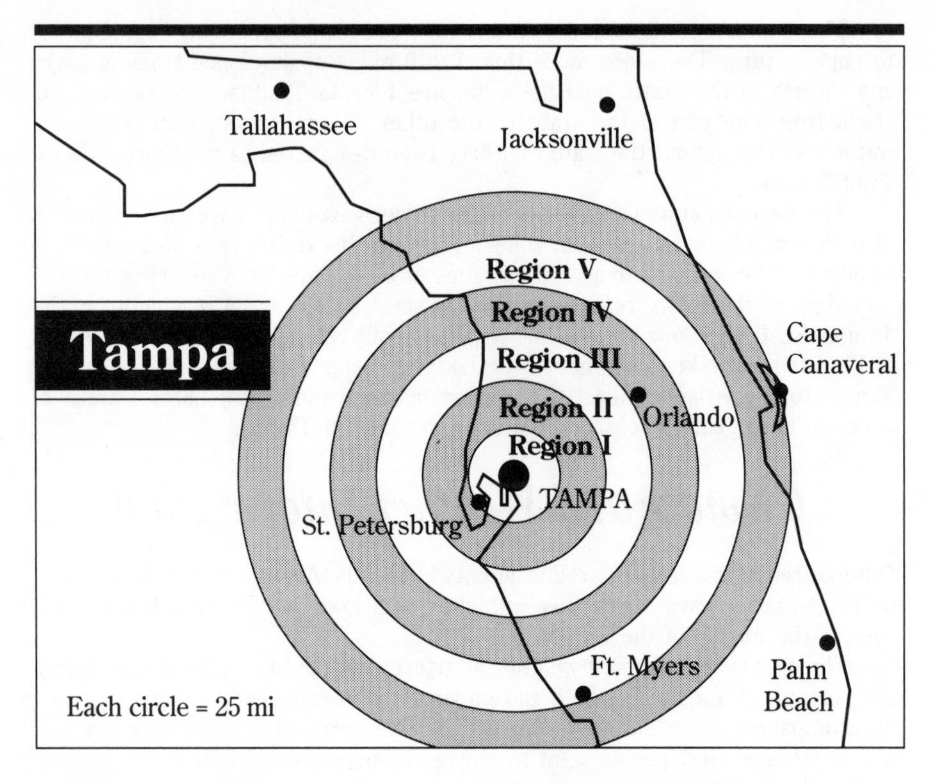

Indoor Activities

Region I

Science Center of Pinellas County
St. Petersburg Junior College
Planetarium and Observatory

Region II

Marie Selby Botanical Gardens
Mote Marine Science Center
South Florida Museum and Bishop
Planetarium

Outdoor Activities

Region I

Caladesi Island State Park
Fort De Soto County Park
Hillsborough River State Park
Lake Maggiore Park
Simmons Park

Region II

Lake Manatee State Recreation Area
Sarasota Jungle Gardens
Suncoast Seabird Sanctuary
Suwanee National Wildlife Refuge
Withlacoochee State Forest

Region III

Chasahowitzka National Wildlife Refuge
Lake Griffin State Recreation Area
Myakka River State Park
Oscar Scherer State Recreation Area

Region IV

Cayo Costa State Park
Highlands Hammock State Park
Lake Kissimmee State Park
Lithia Springs Park
Paynes Prairie State Preserve
San Felasco Hammock State Preserve
Sanibel Island

Region V

Briggs Nature Center
Canaveral National Seashore
Collier-Seminole State Park
Corkscrew Swamp Sanctuary
Devil's Millhopper State Geological Site

Display House and Museum of Botany and Art are both indoor activities that can be enjoyed during inclement weather. The gardens are open daily from 10:00 A.M. to 5:00 P.M.

Mote Marine Science Center
State Route 789, Sarasota, FL 34230—813-388-2451

Sea life native to the Sarasota Bay and the Gulf of Mexico are on display in tanks of varying size. Research projects and a shell exhibit are also found inside. An outdoor tank gives visitors opportunities to view sharks. The center is open Tuesday–Friday 10:00 A.M.–6:00 P.M. and Saturday–Sunday, 10:00 A.M.–4:00 P.M. It is at the south end of New Pass Bridge on State Route 789.

Science Center of Pinellas County, Inc.
7701 Twenty-Second Ave., North, St. Petersburg, FL 33710—813-384-0027

This working lab provides scientific training to students from kindergarten through adults. It is open to visitors from 9:00 A.M. to 4:00 P.M., Monday–Friday, except holidays.

South Florida Museum and Bishop Planetarium
201 Tenth St., West, Bradenton, FL 34205—813-746-4131

A diorama describes the history of Florida from the Stone Age to the Space Age. The museum also has an extensive collection of Indian artifacts. It is open Tuesday–Friday, 10:00 A.M.–5:00 P.M. and Saturday–Sunday, 1:00 P.M.–5:00 P.M. Planetarium shows are daily except Monday at 3:00 P.M.

St. Petersburg Junior College Planetarium and Observatory
6605 Fifth Ave., North, St. Petersburg, FL 33710—813-341-4320

This center, refurbished in 1988, offers star programs and viewings for visitors on Friday evenings from September through April. Contact them for other activities and open hours.

Outdoor Activities

Tampa, and the surrounding areas on Florida's west coast, offer naturalists outstanding opportunities to explore a variety of activities. Most of these can be enjoyed year-round.

Briggs Nature Center
c/o The Nature Conservancy, 11450 Merrihue Dr., Naples, FL 33942—813-262-0304

This is a Nature Conservancy center on Rookery Bay that is an important bird refuge. You will see many of the same birds here as in Collier-Seminole State Park, with fewer people. Bird watching, canoeing, and boardwalk tours are all offered.

Caladesi Island State Park
P.O. Box B, Dunedin, FL 33528—813-443-5903

Located north of Clearwater off the coast of Dunedin, this park is accessible only by boat.The 1,700-acre island has one of the few undisturbed barrier beaches in Florida. The gulf side of island has more than two miles of white sand beach. The bay side is primarily mangrove swamp. The interior is virgin pine flatwoods.

Over 160 species of birds have been recorded on Caladesi Island, but there are few mammals. Rattlesnakes are reported to be big and numerous.

A nature trail extends from the developed area to a water hole in the interior, and guided walks are given according to seasonal demand. A 60-foot observation tower gives a panoramic view of island. The park is open from 8:00 A.M. to sunset.

Canaveral National Seashore
P.O. Box 2583, Titusville, FL 32780—305-867-4675

Located on a barrier island on the Atlantic shore between New Smyrna and Titusville, this 67,500-acre seashore refuge is a two- to three-hour drive from Tampa. It is one of the few remaining wilderness areas on Florida's Atlantic coast and is worth the drive. Legislation that created the seashore refuge directed that land be left natural and primitive; therefore there are few roads on the island. The road on the southern end runs for about five miles behind dunes, and the road on north end dead-ends at the seashore boundary.

Over 700 species of plants and 280 species of birds have been recorded in the seashore. Manatee are often seen in Mosquito Lagoon, a subtropical estuary that lies between the island and the mainland.

This refuge is an excellent wildlife area, and though there are no marked trails, it is exhilarating to hike among the dunes, hammocks, and marshes. The seashore headquarters has a helpful publication entitled *Guide to Coastal Vegetation in Canaveral National Seashore.*

Cayo Costa State Park
P.O. Box 1150, Boca Grande, FL 33921—813-966-3594

This park is accessible only by boat, and occupies the major portion of two islands, Cayo Costa and North Captiva. Cayo Costa is one of the largest undeveloped barrier islands in Florida, and looks much as it did 500 years ago.

Botanically, the island is interesting because of the mix of native plants from both the temperate zone of North America and the West Indian tropics. Bird life is spectacular, with ospreys and bald eagles nesting here, along with the largest known brown pelican rookery in the state. A large colony of white pelicans winters here, and loggerhead sea turtles nest on the beaches in the summer.

A charter boat that makes regular trips to the island operates out of Bokeelia. It is located at the end of Tortuga Street on the north end of Pine Island. There is camping on the island, plus 12 rustic cabins that are available by reservation only.

Chasahowitzka National Wildlife Refuge
Route 2, P.O. Box 44, Homosassa Springs, FL 32646—904-628-2201

Few out-of-state visitors, and only about 35,000 total visitors a year see this 30,000-acre refuge. Most come to hunt and fish. Mazes of marshes, bays, swamps, and mangroves, and teeming wildlife dominate this refuge. A boat or canoe is needed to get into the interior of the refuge where the most wildlife can be spotted. Fortunately water craft can be rented nearby.

Hardwood swamps, hammocks reigned over by cypress, and other southern forest trees dominate the refuge near the headquarters. These give way to tidal bays and winding pathways through large expanses of marsh grass.

Bird life is overwhelming here, with waterfowl, wading birds, and migrant songbirds all represented in the winter. Above all of these soar bald eagles and osprey.

Manatee, or sea cows, can also be observed in the warm waters of the river during the winter. They come up river to avoid the chill of the open sea.

Collier-Seminole State Park

Route 4, P.O. Box 848, Naples, FL 33961—813-394-3397

Most of the 6,500 acres in this park are mangrove swamps. There are also cypress swamps and saltwater marshes, along with a string of hammocks. These are home to one of the largest stands of royal palms in the United States.

A nature trail near the campground leads visitors through a jungle of West Indian hardwoods, but the 13½-mile canoe trail that winds through tidal creeks, bays, and mangroves offers a much better view of the habitats here. Roseate spoonbills, herons, white ibis, and egrets abound in the park, and lesser known species such as the mangrove cuckoo can also be seen. Canoes are limited to 30 per day on the canoe trail, and visitors who wish to be included in that number should call the park headquarters for information about reservations.

Corkscrew Swamp Sanctuary

Route 6, P.O. Box 1875-A, Naples, FL 33942—813-657-3771

This is one of the most famous refuges of the National Audubon Society, as well as one of the most important. It lies northeast of Naples on the northern rim of the Big Cypress, where the piney flatlands of central Florida begin. The sanctuary includes the last great stand of virgin bald cypress in the United States.

Over 10,000 acres in the refuge offer visitors many opportunities to observe alligators, displays of orchids and other air plants, and the secretive brown limpkin, a bird seldom seen elsewhere, but often visible here.

The largest remaining rookery of wood storks in the United States is also located within the refuge. When the water level is right, thousands of these large, striking birds gather to raise their young. Unfortunately, humans have manipulated the water conditions in the region in such a manner that the wood stork is experiencing many breeding failures in recent years.

One of the longest boardwalks anywhere, over two miles long, leads visitors through a portion of the swamp. This gives them a chance to see and feel the many features of a primeval cypress wilderness, while keeping their feet dry.

Devil's Millhopper State Geological Site
4732 NW 53rd Ave., Gainesville, FL 32606—904-336-2008

This site features a sinkhole 120 feet deep and 500 feet across that was formed when an underground cavern collapsed. Plants and animals normally found in ravines in the Appalachian Mountains can be found here. A quarter-mile wooden walkway descends into the hole. Guided tours are offered on Saturdays at 10:00 A.M. There is also an interpretive center that houses exhibits on the natural history of the site.

Fort De Soto County Park
Tierra Verde, FL 33715—813-866-2662

Located south of St. Petersburg on State Route 693, this park is composed of five islands totaling 900 acres that are linked to the mainland by two causeways. Its location at the tip of an island chain, and at the mouth of a bay, makes it one of the best bird-watching spots in Florida. The best time for birds is during the spring migration between late March and late May. Some species winter over.

Sandy beaches, brush, open fields, and pine woods are also featured. The Pinellas National Wildlife Refuge adjoins the park, and the Egmont Key National Wildlife Refuge, an inaccessible island, is nearby.

Highlands Hammock State Park
Route 1, P.O. Box 310, Sebring, FL 33870—813-385-0011

Located six miles west of Sebring, this 3,800-acre park was one of Florida's four original state parks. It was established to preserve a virgin hardwood forest. The park includes several other plant communities in addition to the hammock, and a variety of wildlife. Alligators can be seen daily. Otters and Florida scrub jays are common. Bald eagles and Florida panthers are occasionally sited. There is a ranger-led tram tour to remote areas of the park, and eight separate nature trails, including a catwalk through a cypress swamp.

Hillsborough River State Park
15402 U.S. Route 301 North, Thonotosassa, FL 33592—813-986-1020

This park is northeast of Tampa on U.S. Route 301. The river flows over limestone outcrops and through swamps and hammocks with bald cypress, magnolia, live oak, and sabal palm in its 3,000 acres. Over 150 species of birds have been reported here. Alligators are frequently seen, and many other mammals and reptiles can be observed.

A suspension bridge crosses the river to the nature trails that meander through the hammocks and along the river. Guided walks and camp fires are given at various times during the year.

Lake Griffin State Recreation Area
103 Highway 441-27, Fruitland, FL 32731—904-787-7402

This 427-acre park sits in terrain that is typical of Florida's Central Ridge. The uplands of the park are separated from Lake Griffin by an extensive marsh where the plant life actually floats on a dense mat in several feet of water. This mat consists of soil and roots built up over a number of years. Occasionally a piece of this mat will break off and become a floating island in the lake. A nature trail leads visitors through a live oak hammock and along the marsh.

Lake Kissimmee State Park
14248 Camp Mack Rd., Lake Wales, FL 33853—813-696-1112

This 5,000-acre park is part of the Osceola Plain, and Lake Kissimmee is a prominent feature. Wildlife is abundant. Several pair of bald eagles nest on Buster Island, a slightly elevated sandy area that has scrubby and pine flatwood vegetation circled by a hardwood hammock. Sandhill crane are often seen feeding in the prairie. An observation platform is located at the picnic area, and the 13-mile hiking trail system has two loops, one of which circles a wilderness preserve.

Lake Maggiore Park
Twenty-Second Ave. & Country Club Way, St. Petersburg, FL 33712—813-893-7326

This 700-acre park includes the Boyd Hill Nature Trail, a nature center, and Lake Maggiore. Picknicking, fishing, biking, and boating are all favorite activities in the park.

Lake Manatee State Recreation Area
20007 State Road 64, Bradenton, FL 34202—813-746-8042

Located 14 miles east of Bradenton, this 556-acre park extends along the shore of Lake Manatee. It is the home of the gopher tortoise and indigo snake, both threatened species. Many small mammals can be observed in the park. Fire trails lead through the forest and are excellent for hiking.

Lithia Springs Park
Route 3, P.O. Box 438, Lithia, FL 33547—813-689-2139

This 160-acre park centers around Lithia Springs, a large artesian that averages a flow of 24 million gallons a day, and has a year-round temperature of 72°F. The Alafia River meanders through the park, and offers a variety of plant communities and wildlife. There is a short nature trail.

Myakka River State Park
13207 State Road 72, Sarasota, FL 34241—813-924-1027

Follow State Route 72, nine miles east of Interstate 75. This is one of Florida's largest parks with almost 29,000 acres. A 7,500-acre section of the park has been set aside as a wilderness preserve where the number of daily visitors is limited.

Visitors can view an abundant wildlife population in a landscape that has changed little since the first Europeans settled Florida. There are over 40 miles of trails within the park, some self-guided nature trails, and many roads. All of these offer opportunities for visitors to view many species of wading birds. Waterfowl come by the thousands in the winter. Ospreys, bald eagles, and sandhill cranes can also be seen here.

On a seasonal basis, Myakka Wildlife Tours operate a guided air boat tour and a tram safari land tour into remote areas. Call them at 813-365-0100 for information.

Nature Trails

Other parks around Tampa–St. Petersburg that have nature trails and activities include the following.

Crystal River State Archaeological Site
3400 N. Museum Point, Crystal River, FL 32629—904-795-3817

Features include nature trails through a mound complex.

Dade Battlefield Historic Site
Bushnell, FL 33513—904-793-4781

De Leon Springs State Recreation Area
DeLeon Springs, FL 32028—904-985-4212

DeSota National Monument
P.O. Box 15390, 75th St., NW, Bradenton, FL 34280-5390—831-792-0458

Fort Cooper State Park
3100 S. Old Floral City Rd., Inverness, FL 32650—904-726-0315

Honeymoon Island State Recreation Area
State Road 586, Dunedin, FL 34698—813-734-4255

Lake Woodruff National Wildlife Refuge
P.O. Box 488, DeLeon Springs, FL 32028—904-985-4673

Two trails are near the headquarters.

Oscar Scherer State Recreation Area
P.O. Box 398, Osprey, FL 33559—813-966-3154

This 462-acre park consists of pine flatwoods and scrub. It sits on the bank of a tidal creek. It is noted for its Florida scrub jays, which are a threatened species. Bald eagles are commonly seen during the winter. Bobcats, otters, and alligators are all seen occasionally.

Paynes Prairie State Preserve
Route 2, P.O. Box 41, Micanopy, FL 32667—904-466-3397

Located ten miles south of Gainesville on U.S. Route 441, this 18,000-acre preserve includes freshwater marsh, hammocks, pine flatwoods, swamps, and ponds. There is a visitor center and a 50-foot observation tower near the center of the preserve. Interpretive activities are conducted by the staff. The preserve is open daily from 8:00 A.M. to sunset.

San Felasco Hammock State Preserve
4732 Northwest 53rd Ave., Gainesville, FL 32606—904-336-2008

Located northwest of Gainesville on State Route 232, this 6,000-acre preserve offers visitors an opportunity to view both wildlife and geologic formations.

A nature trail leads south of State Route 232 into the preserve. Guided ranger walks are given from October through April.

Sanibel Island

While not all of this island, and its smaller neighbor Captiva, has been declared a natural area, the Sanibel–Captiva Conservation Foundation has spearheaded an effort to maintain the natural qualities of the islands. It operates a Conservation Center on 207 acres with nature trails around the Sanibel River, naturalist's presentations, and guided tours.

Shelling experts have declared Sanibel the best shell-collecting area in the western hemisphere. More than 200 species of birds have been recorded on the island.

About half of the 15,000 acres on the island are publicly owned, and much of the western side is set aside as the J.N. "Ding" Darling National Wildlife Refuge. A one-way road leads through the bayous here, separating the fresh water from the salt. It gives visitors opportunities to view mangrove wetlands dotted with tropical hammocks.

Sarasota Jungle Gardens

3701 Bayshore Rd., Sarasota, FL 34234—813-355-5305

This ten-acre commercial garden and theme park is a good place to take children, and to get a view of many tropical plants and birds. A variety of shows are presented each day.

Simmons Park

P.O. Box 1416, Ruskin, FL 33570—813-645-3836

This 458-acre regional park sits on the shores of Tampa Bay. It has a nature preserve and excellent bird watching on the shallow waters of the mangrove swamps surrounding the park.

Suncoast Seabird Sanctuary

18328 Gulf Blvd., St. Petersburg, FL 33708—813-391-6211

This sanctuary houses and treats injured pelicans, hawks, owls, egrets, herons, and other birds. They are housed until able to fly, then released. Those with permanent injuries remain at the sanctuary. The staff conducts informal tours of the facilities. The sanctuary is open daily from 9:00 A.M. to dusk.

Suwanee National Wildlife Refuge

c/o Chassahowitza National Wildlife Refuge, Route 2, Box 44, Homosassa Springs, FL 32646—904-628-2201

This refuge is hard to locate. You should ask for good directions in Manatee Springs. Suwanee is 50,000 acres of hardwood forests including oak, red maple, ash, and sweet gum that alternate with cypress heads, small ponds, sloughs, and pine forests. Pileated and red-bellied woodpeckers, little blue herons, egrets, cardinals, and nuthatches can all be seen on walks along the many dirt roads that lead into the refuge.

Withlacoochee State Forest
15023 Broad St. Brooksville, FL 33512—904-796-5650

This forest has four separate tracts totaling 113,431 acres extending over 45 miles in a sparsely populated section of west central Florida. Maps and information about all the hiking trails can be obtained from the forest manager.

The Forest Headquarters Tract is on U.S. Route 41 between Brooksville and Inverness. This tract includes the McKethan Lake Recreation Area where a two-mile nature trail circles this small, popular fishing lake. It also includes the Colonel Robins Recreational Area, where there are several nature trails, including one more than three miles long.

The Croom Tract is composed of 20,470 acres where recreation is the primary activity, with game management and timber production secondary. This tract is located on both sides of Interstate 75 near the town of Croom.

Silver Lake, the widest expanse of the Withlacoochee River, is the focus of much recreational activities in this area. There are almost 30 miles of hiking trails here, including the 19-mile Croom Hiking Trail that is a designated section of the Florida Trail. Some of these trails wind through ravines, prairies, and near abandoned rock mines. Another runs along the river from the Hog Island Recreation Area to the River Junction Recreation Area.

In the Richloam Tract, a 42,200-acre section of fertile soil where timber production is the primary activity, hikers can walk along a 25-mile trail that was built by the Division of Forestry in cooperation with the Florida Trail Association. Hammocks of sabal palms, which are also known as cabbage palms, rise above the prairie at several sites here. The sabal is the official state tree of Florida. This tract is located off Interstate 301 near Ridge Manor.

The 42,531-acre Citrus Tract is managed primarily for wildlife, but it also has more than 45 miles of hiking trails in four different loops. This tract, located between State Routes 491 and 581 near Inverness, has a deer herd that has been estimated at more than 2,000.

Organizations That Lead Outings

The Conservation Center
Sanibel–Captiva Conservation Foundation, 3333 Sanibel–Captiva Rd., Sanibel, FL 33957—813-472-2329

Florida Trail Association
P.O. Box 13708, Gainesville, FL 32604—904-378-8823

Myakka Wildlife Tours
13207 State Road 72, Sarasota, FL 34241—813-365-0100

National Audubon Society
410 Ware Blvd., Tampa, FL 33619—813-623-6826

A Special Outing

Thirty-five primitive rivers and waterways have been designated as the Florida Canoe Trail System. At least seven of these—Econlockhatchee, Withlacoochee, Pithlachascotee, Alafia, Little Manatee, Upper Manatee, and Peace Rivers—are within an hour's drive of Tampa–St. Petersburg. While this system is far from the white water runs that many northern rivers offer, its does give naturalists opportunities to explore wilderness areas of Florida that are inaccessible on foot or by car. Even inexperienced canoeists can handle the slow-moving waterways of central Florida.

Canoe rental facilities are located all over the state near lakes and rivers. The *Guide to Florida Canoeing Rivers* includes a list of 35 canoe rental agencies in the state, some of which provide shuttle service and complete outfitting. This book and other information on the trails can be obtained from the Department of Natural Resources, Bureau of Education and Information, Crown Building, 3900 Commonwealth Boulevard, Tallahassee, FL 32303; 904-488-7326.

The U.S. Forest Service provides *Canoeing the National Forests in Florida*, and it can be obtained from the U.S. Forest Service, 227 North Bronough Street, Suite 4061, Tallahassee, FL 32301; 904-681-7265. The Suwannee River Management District, Route 3, P.O. Box 64, White Springs, FL 32060; 904-362-1001, provides the *Suwannee River Canoe Trail Guide*. Several publications that provide maps and details of canoeing in the Everglades can be obtained from the Everglades National Park, P.O. Box 279, Homestead, FL 33030; 305-247-6211.

There are also two highly recommended books on canoeing in Florida. The first is *A Canoeing and Kayaking Guide to the Streams of Florida* by Elizabeth Carter and John Pearce, and *A Guide to Wilderness Waterways* by William Truesdell. The first can be purchased at local bookstores, and the latter from the Everglades National Park.

The trails around Tampa range in length from five to 84 miles, and wind through some of the prime wilderness area of central Florida. These routes are along the Pithlachascotee, Little Manatee, and Upper Manatee Rivers. The five-mile routes near Tampa are relaxed half-day trips, and are easy to canoe both ways. This means you can make the trip with only one car. These outings take you away from the crowds, are particularly pleasant from late fall through winter, and give you an unforgettable naturalist experience.

Nature Information

Department of Natural Resources
Division of Recreation and Parks, Crown Bldg., 3900 Commonwealth Blvd., Rm. 613, Tallahassee, FL 32303—904-488-6131

Division of Forestry, Department of Natural Resources
State Capitol, Tallahassee, FL 32304—904-488-3022

Office of Visitor Inquiry
Florida Division of Tourism, 126 Van Buren St., Tallahassee, FL 32301—904-487-1462

St. Petersburg Area Chamber of Commerce
401 3rd Ave., South, St. Petersburg, FL 33701—813-821-4069

Tampa–Hillsborough Convention and Visitors Association
100 S. Ashley Dr., Suite 850, Tampa, FL 33602—813-223-1111, 800-826-8358

Further Reading

Boyce, Chris. *Florida Parks Guide.* Wauwatosa WI: Affordable Adventure, 1988.

Bowman, Margaret C., and Herbert W. Kale, II. eds. *Where to Find Birds in Florida.* Maitland, FL: Florida Audubon Society, 1977.

Carter, Elizabeth F.A. *A Hiking Guide to the Trails of Florida.* Hillsborough NC: Menasha Ridge Press, 1988.

Fichter, George S. *Birds of Florida.* Miami: Seemann Publishing, 1971.

Firestone, Linda, and Whit Morse. *Florida's Enchanted Islands: Sanibel and Captiva.* Chesterfield VA: Good Life Publishers, 1980.

Fleming, Glenn, Pierre Genelle, and Robert W. Long. *Wild Flowers of Florida.* Miami: Banyan Books, 1976.

Grow, Gerald. *Florida Parks: A Guide to Camping With Nature.* Tallahassee, FL: Longleaf Publishing, 1987.

Lakela, Olg, and Robert W. Long. *Ferns of Florida.* Miami: Banyan Books, 1976.

Toner, Mike and Pat. *Florida by Paddle and Pack.* Miami: Banyan Books, 1979.

Walden, Fred. *A Dictionary of Trees: Florida and Sub-Tropical.* St. Petersburg, FL: Great Outdoors Publishing Co., 1963.

Memphis

Tennessee is noted for two outstanding recreation areas. The first is the mountain area of the east, including The Great Smoky Mountain National Park, and the second is the system of reservoirs built by the Tennessee Valley Authority in the central portion of the state. Memphis, which sits on the Mississippi River along the western border of the state, is about half a day's drive from either of these.

Tennessee, which is 432-miles long, is separated into three distinct geographical regions. The Great Smoky Mountains of the east rise to over 6,000 feet, and spread westward to the lower Cumberland Plateau. This plateau drops abruptly into the Highland Rim Plateau of central Tennessee. This region reaches nearly the width of the state, and includes the Bluegrass area.

ABOVE: Visitors to Memphis' Mud Island can walk along a flowing scale model of the lower Mississippi River. Each step equals a mile, so it is a 20-minute stroll "upriver" from the one-acre replica of the Gulf of Mexico to Cairo, Illinois. Photo by Allen Mims.

The Tennessee River is the boundary between the central and western regions of the state. From there the state slopes gently to the west until shortly before the Mississippi River. There a sharp bluff stands above the swampy bottomland that lies beside the mighty Mississippi. In most respects the land of extreme western Tennessee is a coastal plain, not unlike southern Louisiana or the coast of the Carolinas.

Memphis sits in the extreme southwest corner of the state, and near the borders of Mississippi and Arkansas. Though not blessed with as many surrounding natural areas as some of the cities in this guide, Memphis has developed a fine city park system. Visitors enjoy natural outings along the Mississippi River and in Arkansas and Mississippi.

Climate and Weather Information

Memphis has a temperate climate, though its summers can be hot and humid with high temperatures in the 80s and 90s and with humidity readings to match. Winters average a high of 42°F, but there are days when the temperature drops lower, and a chill drives residents indoors. Most rain falls during spring and summer.

Getting Around Memphis

Memphis has been much more conservative in its approach to growth than the boom towns of Atlanta and Houston. As a result, it has been able to develop its freeway system in a controlled manner. This makes driving in the city easy for visitors. That is good, for almost all of the naturalist activities included here are away from the city proper, and require that you rent a car.

Rush hours, as with all major cities, should be avoided if possible. Finding your way around town is relatively easy since Memphis' street plan is developed with the river as its focus. All streets run either parallel or perpendicular to it.

Public transportation is adequate in the city, and taxi fares are reasonable. Parking is at a premium in downtown and near major attractions.

Indoor Activities

These activities can be enjoyed during the hot days of summer. They are also appropriate during those times during the winter when the wind and wet chills visitors.

Chucalissa Indian Museum
Mitchell Rd., Memphis, TN 38109—901-785-3160

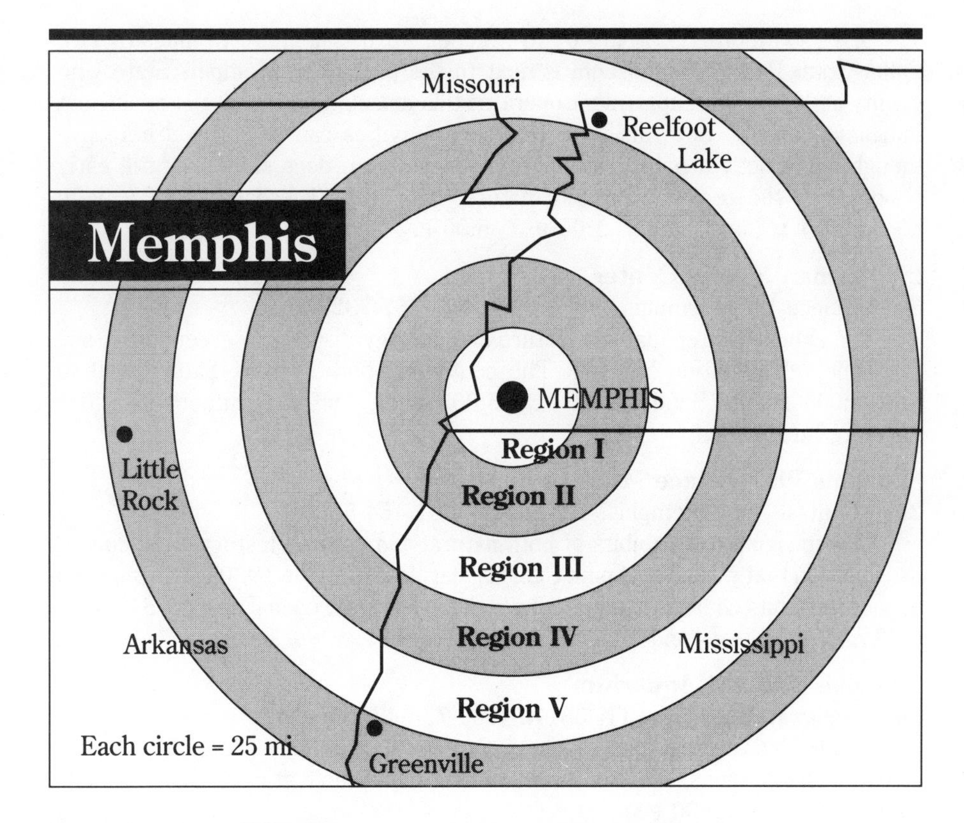

Indoor Activities

Region I
Chucalissa Indian Museum
Lichterman Nature Center
Memphis Pink Palace Museum
Memphis Zoo and Aquarium
Mud Island

Outdoor Activities

Region I
Dixon Gallery and Gardens
Lichterman Nature Center
Meeman-Shelby State Park
Memphis Botanic Garden
Overton Park
T.O. Fuller State Park

Region II
Chewalla Lake and Recreation Area

Region III
Hatchie National Wildlife Refuge

Region IV
Big Cypress Tree State Natural Area
Crowley's Ridge State Park
Reelfoot National Wildlife Refuge

Region V
Arkansas Post National Memorial
Blanchard Springs Caverns
Great River Road State Park
Louisiana Purchase State Park
Tennessee National Wildlife Refuge

Drive south on U.S. Route 61 to Mitchell Road. Then go five miles to T.O. Fuller State Park. The museum is next to the park. The Memphis State University anthropology department operates this museum on the site of an ancient Indian settlement. Both inside and outside activities can be enjoyed here. Although not strictly a naturalist activity, the museum does show how the early residents of the region lived with nature. It is open Tuesday–Saturday, 9:00 A.M.–5:00 P.M., and Sunday, 1:00 P.M.–5:00 P.M.

Lichterman Nature Center
5920 Quince Rd., Memphis, TN 38119—901-767-7322

The exhibit center features nature and ecology displays, a greenhouse, and a wildlife rehabilitation building. The center is open Tuesday–Saturday, 9:30 A.M.–5:00 P.M., and Sunday, 1:00 P.M.–5:00 P.M. See entry in Outdoor Activities for more information.

Memphis Pink Palace Museum
3050 Central Ave., Memphis, TN 38111—901-454-5600

This museum has exhibits of both natural and cultural history. It is housed in a mansion that was built of pink Georgia marble during the 1920s. The museum includes exhibits on birds, fossils and a planetarium. It is open Tuesday–Saturday, 9:30 A.M.–4:00 P.M., and Sunday, 1:00 P.M.–5:00 P.M.

Memphis Zoo and Aquarium
2000 Gallaway, Memphis, TN 38112—901-726-4775

Located in Overton Park, just off Popular Avenue, the Memphis Zoo and Aquarium has regional exhibits as well as traditional zoo features. It is open daily from 9:00 A.M. to 5:00 P.M.

Mud Island
125 N. Front St., Memphis, TN 38103—901-576-7241

This 50-acre island in the Mississippi River near downtown presents a variety of aspects of the river. Included are a five-block long scale model of the lower Mississippi and a five-story river museum with 18 galleries. From May to October, the museum is open daily from 10:00 A.M. to 10:00 P.M. Call the museum for available hours during the rest of the year.

Outdoor Activities

Naturalist acitivities in and around Memphis give visitors a chance to explore some of the bottomlands that once covered most of western Tennessee, Arkansas, and Mississippi. Some of these are included in the following list.

Arkansas Post National Memorial
Gillett, AR 72055—501-548-2432

Located south of Memphis, off U.S. Route 165, this site is considered the birthplace of Arkansas. The post was first established in 1686 by Henri de Tonti.

During the 18th and 19th centuries the town site was relocated many times within a 35-mile area because of general instability of the river's path, and the consequent floods.

Several of these sites are within today's park where self-guiding trails take visitors through 221 acres of wildlife preserve at the meeting of the White, Arkansas, and Mississippi rivers. John James Audubon is reported to have bird-watched in the area, and bird watching is still excellent.

Big Cypress Tree State Natural Area
Kimery Rd., Greenfield, TN 38230—901-235-2700

Located northeast of Memphis off Interstate 40 on U.S. Route 45E, the center of this nature preserve is a giant 18-foot cypress tree, one of many that once covered the surrounding bottomland. The hardwood forest is on the flood-plain of the Middle Fork of the Obion River. The natural area has an interpretive center and a wooden walkway that takes visitors through the bottomland.

Blanchard Springs Caverns
Mountain View, AR 72560—501-757-2213

Take U.S. Route 63 west from Memphis to State Route 14. Follow the state road to Mountain View, Arkansas. These caverns are not as well known as New Mexico's Carlsbad Caverns, but they are as spectacular in many ways. Visitors can take the easy Dripstone Trail year-round. The Discovery Trail, which takes an hour and 40 minutes, includes 700 steps, is available from April through October. The caverns are found in Ozark National Forest near a white-water stream. An adjoining recreation area offers camping and trails.

Chewalla Lake and Recreation Area
c/o U.S. Forest Service, 100 W. Capitol St., Suite 1141, Jackson, MS 39269—601-960-4391

Although fishing and swimming are the primary activities at this recreation area, there are several nature trails. It is surrounded by the Holly Springs National Forest in Holly Springs, Mississippi.

Crowley's Ridge State Park
P.O. Box 97, Walcott, AR, 72474-0097—501-573-6751

This ridge is an erosional remnant left when the Mississippi River moved west at the end of the last Ice Age. It rises 100 to 200 feet above the surrounding flatlands, is from one to ten miles wide, and runs for over 200 miles through Arkansas and Missouri. The park is located on 270 acres along one slope of the ridge where Quapaw Indians once camped.

In addition to the geological oddity of the ridge, a fishing lake is the seasonal home to a variety of waterfowl, including an occasional whistling swan. A network of trails that leads visitors through the surrounding forest is the primary attraction of the park.

Dixon Gallery and Gardens
4339 Park Ave., Memphis, TN 38117—901-741-5250

The fine arts museum is the focal point here, but the 17-acre garden does include both formal and informal gardens, plus some undisturbed woodlands. This is a good nature outing in downtown Memphis.

Great River Road State Park
Rosedale, MS 38769—601-759-6762

Take U.S. Route 61 south from Memphis to Mississippi State Route 1 which is also called The Great River Road. Continue to Rosedale; the park lies between the river and the levee.

Perry Martin Lake, the center of this 200-acre park, is an oxbow of the Mississippi River. A 75-foot observation tower offers a good view of the river, its sandbars, and the cottonwoods and scrubland that lie between the park and the river. The Deer Meadow Trail leads you through pecan and mulberry trees covered with Virginia Creeper. The trail ends in scrubby pastureland. Many birds and deer are generally in sight. The park is open daily from April 15 through Labor Day. During the rest of year, it is open from Wednesday through Sunday.

Hatchie National Wildlife Refuge
34 N. Lafayette St., P.O. Box 187, Brownsville, TN 38102—901-772-0501

Northeast of Memphis, this refuge is ten miles south of Brownsville off Interstate 40. With more than 9,000 acres of timbered bottomland this refuge provides home and shelter for wood ducks, migrating and wintering waterfowl, raccoons, deer, and turkeys.

Lichterman Nature Center
5992 Quince Rd., Memphis, TN 38119—901-726-4775

This is a 65-acre wildlife sanctuary and environmental education center. In addition to the exhibit center, greenhouse, and wildlife rehabilitation facility, there are nature trails for exploring local habitats.

Louisiana Purchase State Park
c/o Arkansas State Parks, 1 Capital Mall, Little Rock, AR 72201—501-682-7777

It was at this spot in 1815, 12 years after the signing of the Louisiana Purchase, that government surveyors blazed two huge sweet gum trees. They marked the starting of the task to determine the boundaries of the new addition to the United States. The blazed trees are gone, but 37½ acres of headwater swamp are now a state park, and a 950-foot boardwalk leads to a marker that designates the initial point of the survey.

The walk leads you through a remnant of swampland which resembles what once covered much of the lower Mississippi Valley. You can see the swamp cottonwood, the golden prothonotary warbler, and the little brown-and-green tree frog. Mosquitos can be a problem, so bring repellent.

The park is on State Route 49 between Binkley and Helena, Arkansas.

Meeman-Shelby State Park
Millington, TN 38053—901-876-5201

Just 16 miles north of Memphis off U.S. Route 51, this 12,512-acre park has many recreational facilities, including hiking trails. Since this is a popular park, especially during spring and summer, it is often crowded and noisy.

An excellent portion of the park to visit is the 9,000-acre bottomland hardwood forest that is home to wild turkeys, deer, bobcats, and bobolinks. Miles of trails lead across river bottoms, around two lakes, and through lush hardwood forests to the banks of the Mississippi. There is also a nature center with exhibits.

Memphis Botanic Garden in Audubon Park
750 Cherry Road, Memphis, TN 38117—901-685-1566

An arboretum, Japanese garden, and collections of iris, cacti, and daylilies are all found in this 100-acre garden. A conservatory displays exotic plants.

Overton Park
2000 Gallaway Ave., Memphis, TN 38112—901-726-4725

With 350 acres, this is the largest of Memphis city parks. It includes the zoo and aquarium, as well as many walking trails. The bird life here is notable.

Reelfoot National Wildlife Refuge
P.O. Box 295, Sambura, TN 38254—901-538-2481

This park is north of Memphis off U.S. Route 51. It includes a 20-mile long lake, a vacation resort, two national wildlife refuges, a state park, and a museum with excellent wildlife exhibits. This is one of the must-see sites near Memphis. The lake was formed during the great Madrid, Missouri earthquake in 1812. It has been famous for its wildlife since at least the time of Davy Crockett.

T.O. Fuller State Park
3269 Boxtown Rd., Memphis, TN 38109—901-785-3950

Southwest of the city limits, off U.S. Route 61, this 1,000-acre park is primarily a golf, swimming, and picnicking area. The Honeysuckle Nature Trail leads visitors through woods populated with deer and wild turkeys. The park is used heavily during the summer.

Tennessee National Wildlife Refuge
P.O. Box 849, Paris, TN 38242—901-642-2091

This 51,347-acre refuge is the wintering area for thousands of waterfowl. A farming program provides supplemental winter food, and attracts a variety of waterfowl that use the Mississippi Flyway.

Organizations That Lead Outings

South Eastern Regional Representative
Sierra Club, S.E. Office, P.O. Box 11248, Knoxville, TN 37939-1248—615-588-1892

Tennessee Environmental Council
1720 West End Ave., Suite 300, Nashville, TN 37203—615-321-5075

Tennessee Native Plants Society
c/o Department of Botany, University of Tennessee, Knoxville, TN 37916—800-341-1000

Tennessee Scenic Rivers Association
P.O. Box 3104, Nashville, TN 37219
and
Tennessee Trails Association
P.O. Box 4913, Chattanooga, TN 37405

No phone queries. Send a self-addressed stamped envelope for current activities.

A Special Outing

In 1908 two promoters planned to develop Reelfoot Lake. Fishermen who used the lake became irate and attacked them during a confrontation. One promoter escaped unscathed; the other was lynched.

While these tactics make even the Earth First! movement seem tame, they only delayed development around this popular lake. Today a resort inn and restaurant are built out over the water's edge among the cypress trees. Vacation cabins, scenic cruise boats, a store, museum, and auditorium are all part of the state resort park that sits on the lake, and attracts thousands of tourists each year.

While this section of the shoreline has a definite tourist atmosphere, other parts of the 14,500-acre lake remain much as it was during the time Davy Crockett hunted bear along the shore. Two national wildlife refuges, the Reelfoot and Lake Isom, in combination with the state-owned lands, form what is often called the greatest hunting and fishing preserve in the nation. They offer some of the best bird-watching opportunities in the United States. Almost every shore and wading bird native to the United States can be found here, as well as bald and golden eagles.

Bald cypress trees surround most of the lake, which was formed as the Mississippi River filled the ten-foot depression left from the Madrid earthquake of 1812. The shallow lake that resulted is dotted with islands, and its partially submerged forest is one of the world's largest natural fish hatcheries. It is home to almost 60 species of fish.

The varied habitat around the lake is also home to some 240 bird species, which makes it one of the best songbird areas of the country. Not only are hunters, fisherman, and bird-watchers attracted to Reelfoot, but the lake's many species of flowering and nonflowering plants bring amateur and professional botanists from around the country.

Winter brings over 250,000 ducks and 50,000 geese to the lake, where they find readily accessible food near the ice-free waters. This season also brings the annual migration of bird-watchers who come to the lake not only to see the waterfowl, but also to see the highest concentration of bald eagles east of the Mississippi River.

As many as 200 of these majestic birds have been observed in a winter at Reelfoot Lake. Between December 1 and mid-March the park and refuge staff conduct eagle tours daily. Visitors are encouraged to utilize this service rather than search for the birds on their own so the eagles aren't disturbed unnecessarily.

Reelfoot Lake is a must for any naturalist who is visiting Memphis. To find out more about eagle tours and other activities, call the refuge headquarters at 901-538-2481, or the state park headquarters at 901-253-7756.

Nature Information

Tennessee Department of Conservation
Division of State Parks, 701 Broadway, Nashville, TN 37203—615-742-6666

Tennessee Wildlife Resources Agency
Ellington Center, Nashville, TN 37220—615-781-6500

Tennessee Department of Tourist Development
P.O. Box 23170, Nashville, TN 37202—615-741-2158

Convention and Visitors Bureau
50 N. Front St., Suite 450, Memphis, TN 38103—901-576-8181

Memphis Area Chamber of Commerce
P.O. Box 224, 555 Beale St., Memphis, TN 38101—901-523-2322

Further Reading

Smith, Arlo. *A Guide to Wildflowers of the Mid-South.* Memphis: Memphis State University Press, 1980.

New Orleans

There is not much variety to the landscape in and around New Orleans. Swamp is what you see and experience. Louisiana has over five million acres of swamp and marsh land, reputed to be more than the other 49 states combined. New Orleans has its share of what the residents call *bayous.*

Just a few miles outside the city, the streets end in exasperation, defeated by wetlands of bayous, lakes, and swamps. These are part of the final stretch of the Mighty Mississippi before it makes its confused and uncertain entry into the Gulf of Mexico. The 80 miles of land, if this mixture of water and delta can be called land, between New Orleans and the Gulf is sparsely populated. Few visitors to the city ever make it to the end of State Route 23, where the river empties into the Gulf.

To the west of New Orleans lies Louisiana's Empty Quarter, where the Atchafalaya River, the largest of the Mississippi's distributaries, drains the 2,000-

---

ABOVE: The Baratana Unit of the Jean Lafitte National Historic Park and Preserve, located just south of New Orleans, has several nature trails. This boardwalk is along the Bayou Coquille Trail. Photo by National Park Service.

square mile Atchafalaya Basin. This is, with its vast areas of bayous, lakes, and cypress swamps, home to an abundance and diversity of wildlife that surpasses even the Florida Everglades. Little of this region has been set aside as park or refuge, but it is still a wild and untamed land that defies human efforts to subdue it.

East of New Orleans, it is a short drive into Mississippi and the barrier islands that separate the Mississippi Sound from the Gulf of Mexico. Along the coast from Biloxi to the Louisiana border are a number of parks and refuges worth visiting.

There is little of the natural world left within New Orleans' city limits. Therefore, naturalists must venture out into the swamps and bayous of southern Louisiana to find activities to satisfy their needs.

Climate and Weather Information

High heat and humidity, along with thick swarms of biting bugs, make summer the most uncomfortable season to explore the outdoors in and around New Orleans. Spring brings high waters to most of the bayous and marshes, making it difficult to enjoy this strange world.

Falls, however, are very pleasant, with cooler and drier days that are not host to multitudes of insects. Winters are normally mild, although these can be wet, windy, and chilly. Be prepared for rain almost any time of year, and bring lots of insect repellent if you venture into the hinterlands.

Getting Around New Orleans

A car is a necessity if you plan to enjoy the outdoors in New Orleans, for many of the attractions listed in this guide are a considerable distance from downtown. Only a few can be reached by public transit, although New Orleans has a good transit system for points of interest within the city. Taxis are plentiful, and rental car agencies can be found at most major downtown hotels.

Indoor Activities

Indoor activities around southern Louisiana are probably more varied than outdoor ones. Some of them are included in the following list.

Acadiana Park Nature Station
E. Alexandra St., Lafayette, LA 70501—318-235-6181

This interpretive center is in a three-story cypress pole structure that helps children and adults get acquainted with the flora and fauna of the adjoining three-

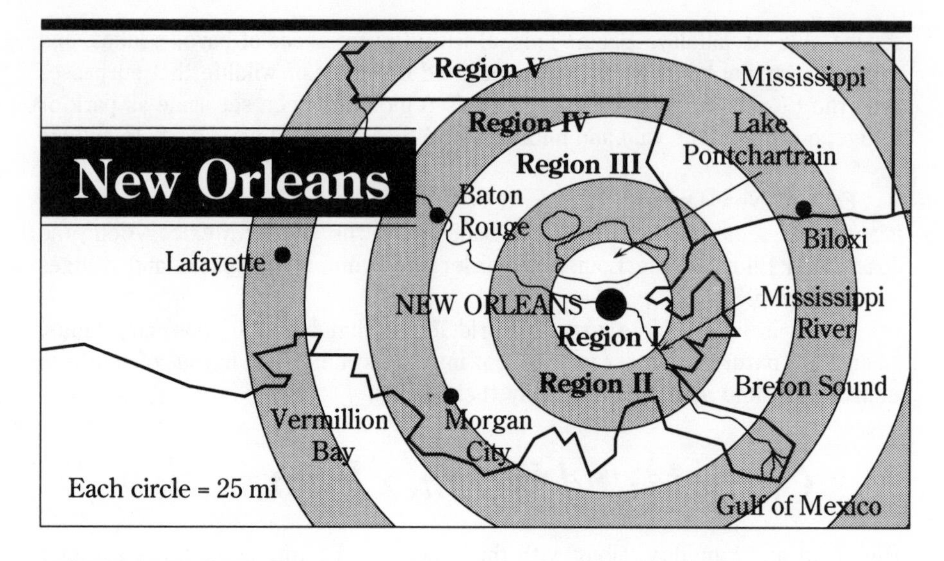

Indoor Activities

Region I

Louisiana Nature Center
Louisiana Wildlife and Fisheries Museum

Region III

Geoscience Museum
Louisiana Arts and Science Center and Riverside Museum
Lyle S. St. Amant Marine Laboratory
Marine Education Center
Museum of Natural Science

Region V

Acadiana Park Nature Station
Lafayette Natural History Museum

Outdoor Activities

Region I

Audubon Park Zoo
City Park
Fairview-Riverside State Park
Fontainbleau State Park
Lake Pontchartrain
St. Bernard State Park

Region II

Amite River
Bogue Chitto National Wildlife Refuge
Brownell Memorial Park
Honey Island Swamp Nature Trail
Jean LaFitte National Park and Preserve
Wildlife Management Areas (see entry)

Region III

De Soto National Forest
Delta-Breton National Wildlife Refuge
Gulf Islands National Seashore
Gulf Marine State Park
Lake Vista Nature Preserve Trail
Laurens Henry Cohn, Sr. Memorial Plant Arboretum
Swamp Gardens
Wildlife Management Areas (see entry)

Region IV

Jungle Gardens
Lake Fausse Point
Mississippi Sandhill Crane National Wildlife Refuge
Wildlife Management Area (see entry)

Region V

Acadian Park Nature Trail
Cypremort Point State Park
Cypress Lake
Marsh Island Wildlife Refuge and Game Preserve

and-one-half-mile trail. The center is open weekdays, 9:00 A.M.–5:00 P.M. and weekends, 11:00 A.M.–3:00 P.M.

Geoscience Museum
Louisiana State University, Geology Bldg., Baton Rouge, LA 70803—504-388-2931

This museum features geological and geographical exhibits of the Mississippi River, its distributaries, and the delta region of southern Louisiana. It is open weekdays from 8:00 A.M. to 4:30 P.M.

Lafayette Natural History Museum
637 Girard Park Dr., Lafayette, LA 70503—318-268-5544

The museum houses many exhibits on natural history, has a nature trail, and features the annual Louisiana Native Crafts Festival in September. It is open Monday–Friday, 9:00 A.M.–5:00 P.M., and Saturday–Sunday, 1:00 P.M.–5:00 P.M.

Louisiana Arts and Science Center and Riverside Museum
100 S. River Rd., Baton Rouge, LA 70802—504-344-9463

Among its other exhibits is a three-dimensional display of channel changes in the Mississippi River. The center also has an indoor overlook of the river. It is open Tuesday–Saturday, 10:00 A.M.–5:00 P.M., and Sunday, 1:00 P.M.–5:00 P.M.

Louisiana Nature Center in Joe Brown Park
11000 Lake Forest, New Orleans, LA 70128—504-246-9381

This 86-acre wilderness park is in the middle of suburban New Orleans. It is devoted to preserving what little is left of the natural environs that once surrounded the city. In addition to a visitors center with exhibits, it has nature trails and offers guided canoe and field trips. The center is open Tuesday–Friday, 9:00 A.M.–5:00 P.M., and Saturday–Sunday, noon–5:00 P.M.

Louisiana Wildlife and Fisheries Museum
New Orleans Court Bldg., 1909 Royal Street, New Orleans, LA 70116—504-568-5667

Numerous native wildlife mounted exhibits, including one of a passenger pigeon, are displayed here. The museum is open Monday–Friday from 8:30 A.M. to 3:45 P.M.

Lyle S. St. Amant Marine Laboratory
Grand Isle, LA 70358—504-787-2559

This laboratory, located at the terminus of State Route 1 on Grand Terre Island, is only accessible by boat. It features exhibits of local marine life. It is part of the Fort Livingston State Commemorative Area that is under development. Exploration of the site is permitted during daylight hours. The laboratory is normally open Monday–Friday, 9:00 A.M.–5:00 P.M.

Marine Education Center
650 E. Beach Blvd., Biloxi, MS 39564—601-374-5550

The center features live exhibits of Gulf Coast animals in 26 small aquariums and a 40,000-gallon aquarium. It is open Monday–Saturday from 9:00 A.M. to 4:00 P.M.

Museum of Natural Science

Louisiana State University, Room 127, Foster Hall, Baton Rouge, LA 70803—504-388-2855

This museum displays life-size dioramas of wildlife scenes, plus mounted exhibits of birds, reptiles, and insects of Louisiana. It is open weekdays, 8:00 A.M.–4:30 P.M. and weekends, noon–4:30 P.M.

Outdoor Activities

If you do not like water and the types of naturalist activities associated with it, you probably will not appreciate most outdoor activities in and around New Orleans. On the other hand, for those who like water activities, and the flora and fauna associated with large bodies of water, the region is paradise.

Acadian Park Nature Trail

E. Alexander St., Lafayette, LA 70501—318-261-8388

This 117-acre park has three and one-half miles of nature trails, an archaeological site, and an interpretive center. It is open daily from dawn to dusk.

Amite River

This clear and shallow river is popular as a canoeing site, and its tributaries provide shelter for large beaver colonies. Beavers were endangered in the area until the 1930s and 1940s, when this area had some of the few flourishing colonies in the South. The river is north of New Orleans off State Route 38 near Chipola, Louisiana.

Audubon Park Zoo

6500 Magazine St., New Orleans, LA 70118—504-861-2537

The zoo includes 58 acres of Audubon Park, and is divided into sections, each representing various habitats. One of these is the Louisiana Swamp exhibit, where native animals including alligators, bask in a miniature bayou. The zoo is open daily from 9:30 A.M. to 6:00 P.M.

Birding Tour

State Route 3090 leads to the nation's only super port, the Louisiana Offshore Oil Port. The marshes and open water along the route provide prime habitat for birds. Many migratory and resident birds can be seen from the highway.

Bogue Chitto National Wildlife Refuge

1010 Gause Blvd., Building 936, Slidell, LA 70458—504-646-7555

This refuge is off State Route 41 near the Mississippi border. Some of the last bottomland hardwood in the United States is saved in this 18,000-acre tract. The unique habitat is active with wildlife year-round. The Pearl River flows through this refuge. Therefore, the canoeing is excellent, giving visitors opportunities to investigate the wildlife away from crowds.

Brownell Memorial Park
725 Myrtle St., Morgan City, LA 70380—504-385-2160

Located north of Morgan City at Lake Palourde, the park is off State Route 70. Native wild plants and huge cypress trees beside the lake are the main attractions here.

City Park
Casino Building, New Orleans, LA 70119—504-482-4888

About three miles from the French Quarter, this 1,500-acre former sugar plantation is one of the largest urban parks in the country. With a network of human-made lagoons and native oak trees, this park is a good place to enjoy nature in the inner city. With an estimated 11 million visitors a year, the park is said to have the highest visitor figure per capita of any urban park in the nation.

Many paths lead through groves of oaks, pines, magnolias, and dogwoods. The formal, seven-acre New Orleans Botanical Garden has pools scattered throughout. In summer it is full of blooms. The park is open daily from sunrise to 10:00 P.M.

Cypremort Point State Park
New Iberia, LA 70560—318-867-4240

State Route 319 leads through cane fields, cypress marshes, and stands of giant oaks on the way to Vermillion Bay. The 185-acre park, 20 miles south of New Iberia, offers access to the Gulf of Mexico and good bird watching.

Cypress Lake
University of Southern Louisiana, University at Johnston, Lafayette, LA 70503—318-233-3850

This is a human-made miniature swamp used as an educational tool. Various kinds of swamp environments, complete with birds and alligators, can be reproduced here.

Delta–Breton National Wildlife Refuge
Venice, LA 70091—504-534-2235

This 70,000-acre tract near the mouth of the Mississippi River offers protection for a variety of migratory birds. A chain of offshore islands in the refuge are accessible only by boat or pontoon plane. This is a shorebird, gull, and tern nesting area. Some of the finest surf fishing in the country is available during daylight hours. The refuge is at the end of State Route 23, south of Venice.

De Soto National Forest
Route 1, P.O. Box 62, McHenry, MS 39561—601-928-5291

Two hiking trails, the 17-mile Tuxachanie and the 40-mile Black Creek, offer some of the most scenic hiking in the South. Canoeing on over 90 miles of Black Creek is one of the main attractions of this national forest.

Southern bogs and piney woods combine with bottomland hardwoods to provide excellent examples of the diverse flora of the Gulf Coast forests. The floors are veretible flower gardens all spring and summer.

Maps and information are available from the above address. The forest is north of Biloxi off State Route 49.

Fairview–Riverside State Park

Madisonville, LA 70447—504-845-3318

South of Interstate 12, this 98-acre park is near the Tchefuncte River and Lake Pontchartrain. River and camping activities are featured here.

Fontainbleau State Park

Mandeville, LA 70448—504-626-8052

This 2,800-acre park on the north shore of Lake Pontchartrain has nature trails and the ruins of a plantation sugar mill. It has good access to the lake shore. Fontainbleau State Park is on U.S. Route 190 southeast of Mandeville.

Gulf Islands National Seashore

4000 Hanley Rd., Ocean Springs, MS 39564—601-875-9057

This seashore is comprised of Ship, Horn, and Petit Bois Islands, and the estuaries and marshes inside these barrier islands. It has a nature trail, picnic facilities, and visitor center. Park rangers provide guided tours of the marshes during the summer, and primitive camping is allowed on Horn Island. The park service provides a boat shuttle to the islands. Private boats have access to other islands, and the *Pan American* makes regular trips to Ship Island during the summer months.

Gulf Marine State Park

1700 E. Beach Blvd., Ocean Springs, MS 39564—601-374-5718

A small park jutting out over the water on wooden decks, this is an excellent spot for fishing and bird watching. Telescopes provide views of Deer Island, 12 miles offshore. The museum has wildlife exhibits. The park is at the foot of the Biloxi–Ocean Springs Bridge.

Honey Island Swamp Nature Trail

c/o Pearl River Management Area, P.O. Box 14526, Baton Rouge, LA 70898—504-863-7042

This trail is off Interstate 59 on the east bank of Pearl River. It is a self-guided trail that provides excellent opportunities for visitors to study the natural environment of the region, as well as indulge in photography and bird watching. Use care on the trail during hunting season, for deer hunters roam the area, and are notorious for shooting at all moving objects. Bring plenty of insect repellent during the spring and summer.

Jean LaFitte National Park and Preserve
U.S. Customs House, Room 206, 423 Canal St., New Orleans, LA 70130—504-348-2923

About 30 minutes southwest of New Orleans in Barataria, this park includes wetlands, a museum, and interpretive walks led by rangers. Its 8,600 acres are a small example of the great marshes and swamps that cover most of southern Louisiana. The park is close enough to New Orleans for visitors to be introduced to the natural wonders of the region.

The visitor center and museum display information about the local ecology and hiking and canoeing trails. Three walking trails in the park that are about five miles combined are in place. More are planned, including a trail on the west side of the park that will wind through the estuarine marshes of Lake Salvador. For canoe enthusiasts there is a trail that follows a bayou canal, and takes five to six hours to paddle. Bird watching is terrific here, although it is even better in the large coastal refuges. Small animals such as mink, nutria, and otters are often seen from the trails.

Jungle Gardens
Avery Island, New Iberia, LA 70560—318-365-8173

This 200-acre garden was created by Edward Avery McIlhenny, maker of Tabasco sauce. It now includes many camellias, azaleas, and irises, which fully bloom in February. It is also a bird sanctuary frequented by thousands of herons and egrets. The Jungle Gardens are open daily from 9:00 A.M. to 5:00 P.M.

Lake Fausse Point
500 Main St., Jeanerette, LA 70544—318-276-4293

This pretty, cypress-filled lake is one of the less heavily used lakes in the region. It has a campground, canal, boat ramp, and a guided nature trail.

Lake Pontchartrain

This 610-square mile lake is the largest saltwater lake in North America. It provides an excellent seafood habitat. There is plenty of opportunity to fish, sail, and swim in the lake. A multi-purpose pedestrian trail begins in Linear Park in Metairie, and extends along the south shore of the lake for ten miles to the St. Charles Parish line.

Lake Vista Nature Preserve Trail
214 Arkansas, Bogalusa, LA 70427—504-732-3666

This is a one-mile hiking and nature trail in Cassidy Park that leads through piney woods and cypress swamps in the heart of the yellow pine belt. This area is dramatically different from the bayous near New Orleans.

Laurens Henry Cohn, Sr. Memorial Plant Arboretum
12056 Foster Rd., Baton Rouge, LA 70811—504-775-1006

More than 120 species of native and exotic trees and shrubs adaptable to the region are grown at this 16-acre garden. It is located on a beautiful rolling terrain.

Marsh Island Wildlife Refuge and Game Preserve
c/o Louisiana Department of Wildlife and Fisheries, P.O. Box 14526, Baton Rouge, LA 70898—504-342-5875

Accessible only by boat between Vermillion Bay and the Gulf of Mexico, this island refuge is primarily grass-covered marshlands with lakes and bayous. It is one of the most important wildlife areas for migratory waterfowl and fur-bearing animals in North America.

Mississippi Sandhill Crane National Wildlife Refuge
P.O. Box 6999, Gautier, MS 39553—601-497-6322

This refuge was established to protect a very endangered subspecies, the Mississippi sandhill crane. Its 17,000 acres are set up for just that, and not for the benefit of visitors. You should not bypass the refuge because of that, however, for the visitor center and short nature trail offer an illustration of the natural history of the savannah-pine flatwoods ecosystem. You might get to see one of the 50 or so surviving cranes, although the chances are better that you will hear their rattling call. To get to the refuge, use Interstate 10 and take the Gautier/Vancleave Exit between Biloxi and Pascagoula.

Pascagoula Wildlife Management Area
Lucedale, MS 39452—601-947-6376

The Nature Conservancy masterminded the citizen campaign in the 1970s that convinced the state of Mississippi to purchase 35,000 acres of bottomland forest located in the Pascagoula River Basin and set it aside as a wildlife preserve. Today this preserve is one of the few large tracts of bottomlands to survive in the central part of the Gulf Coast. It is off State Route 63 northeast of Biloxi.

Swallow-tailed kites, herons, wild turkeys, bobcats, deer, and beavers all live in the diverse habitat offered by the bottomlands. Visitors can explore the area on foot or by canoe. Either way, though, you should check in with one of the two headquarter offices for detailed information on the hazards likely to be encountered on and off the trails. One of these offices is located at Parker Lake, three miles off State Route 63, between Vancleave and Wade. The other is off State Route 26, about six miles west of Lucedale. Both locations have maps and other information.

St. Bernard State Park
P.O. Box 534, Violet, LA 70092—504-682-2101

Located on State Route 39 about 18 miles southeast of New Orleans, this 350-acre park on the Mississippi River has many viewing points of the river. Canoeing and fishing are also featured activities.

Swamp Gardens
Heritage Park, Morgan City, LA 70380—504-384-3343

A living museum, this three-and-one-half-acre natural area has raised pathways wandering through a small swamp, complete with alligators and an 800-

year-old cypress. Guided tours with life-sized dioramas and oral narration help visitors understand the human struggle to survive in the Atchafalaya Swamp.

Wildlife Management Areas
Louisiana Department of Wildlife and Fisheries, P.O. Box 14526, Baton Rouge, LA 70898—504-342-5875

Louisiana has set aside thousands of acres of bayou and marsh land as wildlife management areas. These are used primarily for hunting and fishing. Many of these sites also provide naturalists excellent opportunities to view native wildlife. Some of these wildlife management areas are within a reasonable distance from New Orleans. While some of them have nature trails or foot access, many are accessible only by boat. It would be prudent for those unfamiliar with the region to discuss boat rental and use with local boating or fishing stores prior to any outing. For more information on individual units call them or contact the Louisiana Department of Wildlife and Fisheries.

Atchafalaya Delta Wildlife Management Area—This 6,375-acre tract at the mouth of the Atchafalaya River is accessible by boat from Morgan City, Berwick, and Wax Lake. It is one of the best duck locations in state. Call 504-568-5885 for information.

Attakapas Island Wildlife Management Area—Access to this 25,500-acre swampland tract is from Morgan City, Charenton, Bayou Pigeon Landing, and Myette Point. It features good bird watching and some alligators. Call 318-942-7553 for information.

Biloxi Wildlife Management Area—This is a 40,000-acre marsh along the coast formed by Bayou L'Outre. Access is by boat only, but good bird watching and fishing are featured. Call 504-568-5612 for information.

Bohemia Wildlife Management Area—Foot or boat access to this area is from State Route 39 south of New Orleans. Diversified habitats in the 33,000-acre refuge include saline marsh and tree ridges. It has excellent fishing, picnicking, and bird watching. Phone 504-568-5612 for information.

Bonnet Carre Spillway and Wildlife Management Area—The spillway is an emergency floodway to divert water from the Mississippi River to Lake Pontchartrain. During dry seasons, the hardwood timber and grassy marsh offer visitors a break from the heavy swampland around New Orleans. This area is located off U.S. Route 61, west of New Orleans near Laplace. Call 504-342-5875 for information.

Manchac Wildlife Management Area—This 8,325-acre swampland tract between Lakes Maurepas and Pontchartrain is accessible only by boat. There are plenty of alligators, as well as raccoons, muskrats, and nutria here. Phone 504-568-5612 for information.

Pass a Loutre Wildlife Management Area—Access is by boat from Mississippi River at the public launch in Venice. Floating islands of marsh vegetation in this 66,000-acre tract provide food and cover for wildlife, and bird watching and nature study for visitors. A river levee in the area provides picnicking and camping. Call 504-568-5885 for information.

Pearl River Wildlife Management Area—This 26,716-acre tract is along the Pearl River with access primarily by boat. It features good canoeing, fishing, and bird watching. Call 504-342-5875 for information. Also see Honey Island Swamp Nature Trail in the Outdoor Activities section.

Pointe au Chien Wildlife Management Area—This 29,000-acre tract is off State Route 1 south of Larose. Access to the interior of the refuge is limited to boats. Nature study and bird watching, along with primitive camping, is available on edges of the fresh to brackish marshlands.

Salvador Wildlife Management Area—Accessible by boat from several bayous, this is a 31,000-acre freshwater tract along the north shore of Lake Salvador. It offers good bird watching, particularly waterfowl, and plenty of alligators. Phone 504-568-5885 for information.

Wisner Wildlife Management Area—Access is by boat off State Route 1 west of Grand Isle. Good bird watching is featured on this 21,621-acre saline marsh tract. Phone 504-568-5612 for information.

Organizations That Lead Outings

Unlike many cities, most of the organizations listed for New Orleans and vicinity are for-profit organizations, rather than nonprofit environmental and naturalist ones.

Annie Miller's Terrebonne Swamp and Marsh Tours
Houma, LA 70360—504-879-3934

Atchafalaya Basin Backwater Adventure
Gibson, LA 70356—504-575-2371

Two-hour tours of swamps surrounding Gibson and four-hour tours of the Great Chacahoula Swamp are available.

Honey Island Swamp Tours
106 Holly Ridge Dr., Slidell, LA 70458—504-641-1769

Flatboat tours of one of the country's best-preserved river swamps are led by professional wetland ecologists.

Lafayette Natural History Museum
637 Girard Park Dr., Lafayette, LA 70503—318-261-8350

McGee's Landing
Henderson, LA 70517—318-228-8523

Ninety-minute boat tours into the Atchafalaya Basin are provided.

New Orleans Tours and Convention Services
7801 Edinburgh, New Orleans, LA 70125—504-482-1991

This company leads many tours, including a swamp tour to the Jean LaFitte National Park and Preserve.

Sailfish Shrimp Tour
1500 E. Beach Blvd., Biloxi, MS 39564—601-374-5718

These 80-minute tours let visitors cast nets into waters, and pull in their catch. The crew helps identify the catch.

A Special Outing

No self-respecting naturalist would take a trip to southern Florida and not visit the Everglades. But many visit New Orleans and never venture into Louisiana's equivalent, the Atchafalaya Basin. This 800,000-acre wonderland has hardwood forests, cypress swamps, marshes, and bayous that produce more game and commercial fish than any other natural water system in the United States. It supports half the nation's migratory waterfowl, yields over 23 million pounds of crawfish annually, and has an abundance and diversity of wildlife that exceeds the Everglades.

Noncommercial wildlife such as nutria, raccoons, mink, and muskrats live side-by-side with black bears, cougar, and alligator. It is one of the few places in the South where a large bear population exists.

This 18-mile wide and 130-mile long basin is the largest remaining wetland wilderness in the United States. It is a flood control safety valve of the Mighty Mississippi, as it has been for thousands of years.

The basin is populated by small settlements of Cajuns, but has resisted large-scale invasion by civilization. While human settlement has been unable to tame the Atchafalaya, the U.S. Army Corps of Engineers may. Since the 1920s the Corps, in conjunction with farmers and oil companies, has built a series of dams and canals that is causing the basin to fill with silt much faster than it had in the past. Conservationists are concerned that this may be the end of this great wilderness.

Others, however, note that any attempt to tame such a monstrous waterway as the Mississippi and its distributaries is doomed to failure. They say that while the character of the basin may change as a result of the Corps' actions, nature will win in the end.

Unlike the Everglades, no governmental agency has taken control of large portions of the Atchafalaya. The Louisiana Department of Wildlife and Fisheries operates a number of wildlife management areas within the basin, but there are no state or national parks to protect the natural beauty of the region. More than 80 percent of the land is in private hands, and there is considerable pressure to develop more of the area for timber and farming.

For now, though, the region remains the domain of the fishermen and trappers who have lived there for generations. Only two roads cross the lower portion of the basin. Interstate 10 from Baton Rouge to Lafayette crosses 30 miles of swamp on high concrete pylons. Not a house is in view for the whole stretch.

U.S. Route 90 crosses the basin between Houma and Morgan City as the most southerly route. State Routes 1 and 24 skirt the eastern edge of the basin as they go from Baton Rouge to Houma.

In Morgan City visitors can get an introduction to the Atchafalaya Basin with a tour of the Swamp Gardens. More interested visitors can venture into the basin by departing from the freeways, and driving along the levees and backroads of the region. These skirt both the eastward and westward edges of the swamp.

The most adventurous visitors, though, will want to go to the interior of the basin, and that can only be done by boat. Canoes and small motorboats are the best way to do this exploring, but no one should attempt this without either talking over the route with an experienced guide, or taking one along.

Guided tours are the next best, but there are few regularly conducted wildlife tours into the basin. There are several general tours, and these are listed in the Organizations That Lead Outings section.

Maps of the Atchafalaya Basin are available free from the Office of State Parks, P.O. Drawer 1111, Baton Rouge, LA 70821; 504-342-8111. Additional information about the basin can be obtained from the following individuals and groups.

Dr. Charles Fryling, Jr.
1068 East Lakeview Dr., Baton Rouge, LA 70810

Dr. Fryling occasionally leads tours into the basin and has been exploring the region for 15 years. No telephone queries. Send a self-addressed stamped envelope for information.

Louisiana Wildlife Federation
P.O. Box 16089, Louisiana State University, Baton Rouge, LA 70893—504-344-6707

U.S. Army Corp of Engineers
New Orleans District, P.O. Box 60267, New Orleans, LA 70160
Attn: LMNPDA-A—504-865-1121

U.S. Fish and Wildlife Service
11 E. Main St., Lafayette, LA 70501—318-264-6630

The following places offer canoe rentals and guided trips into the basin.

Pack and Paddle
601 Pinhook Rd. East, Lafayette, LA 70503—318-232-5854

The Backpacker, 3378 Highland Rd. Mall, Baton Rouge, LA 70802—504-924-4754

Nature Information

Baton Rouge Area Convention and Visitors Bureau
Old State Capitol Bldg., Baton Rouge, LA 70804—504-383-1825

Houma Tourist Center
U.S. Route 90 at South St. Charles Street, Houma, LA 70360—504-876-5600

Lafayette Convention and Visitors Bureau
16th St. & Evangeline Thruway, Lafayette, LA 70501—318-232-3808

Louisiana Department of Wildlife and Fisheries
P.O. Box 15570, Baton Rouge, LA 70895—504-342-5868

The Greater New Orleans Tourist and Convention Commission
1520 Sugar Bowl Dr., New Orleans, LA 70112—504-566-5031

Further Reading

Boyer, Chris. *Louisiana Parks Guide.* Wawatosa, WI: Affordable Adventure, 1988.

Muller, Carl, and Brenda. *Explore Louisiana.* New Orleans: Bonjour Books, 1984.

M I D W E S T

Cleveland
Chicago
Minneapolis-St. Paul
St. Louis

The westward expansion of the United States began almost as soon as the first cities of the original 13 colonies were built. Men and women soon started moving beyond the Appalachian Mountains into the flatlands and rolling hills of what is now the Midwest. The lands which are now Ohio, Indiana, Kentucky, Missouri, and Illinois had been leveled by glaciers of the last Ice Age. Later, they were blanketed by vast hardwood forests comparable in size to those that now grow in the world's great rain forests. The hardwood forests grew on some of the richest soil found in North America. The regions were drained by extensive river systems.

If this movement were to occur today, environmentalists would probably be outraged at the destruction of virgin forest lands. Two hundred-foot tall hardwoods that had grown for hundreds of years were mercilessly cut to make room for farms and villages to accommodate frontiersmen and their families.

Nothing is left of these pristine forests to remind you of the priceless heritage that was destroyed in the relentless advance toward the Pacific Coast. What is now found in the Midwest are some of the largest cities and most productive farms in the United States, with few natural areas.

Even the vast lakes and mighty rivers of the region have been altered. The factories that were long dominant in the production of the world's goods almost laid waste to both Lake Erie and the Cuyahoga River. Thankfully, both of these great bodies of water are showing signs of recovery from this earlier desecration.

The Chicago River now flows in the opposite direction than it did at the turn of the century. Engineers reversed the flow to keep sewage from entering Lake Michigan and endangering Chicago's drinking water supply. All along the Mississippi River, from Minneapolis past St. Louis to New Orleans, levees and other flood control efforts have changed the natural flow of this mighty river.

Natural History of the Midwest

Of the four midwestern cities in this guide, only Minneapolis has been able to retain significant natural areas near it. Naturalist activities near the other three are harder to find. The lack of nature guides about the region is another confirmation of this.

Water is, and has been, a predominant feature of the natural history of the region. Cleveland and Chicago are located on Lake Erie and Lake Michigan respectively, and each has a major river flowing through it. Minneapolis and St. Louis both sit beside the mighty Mississippi River.

Minneapolis is also blessed with a large number of natural lakes in and around the city. These were formed as glaciers receded. The other three cities have large human-made lakes nearby.

The vast forests that once covered the Midwest are now nearly gone, and naturalists can only guess how the region must have appeared 200 years ago. The same is true for the tall grass prairies that once covered much of Illinois and Minnesota. Only small samples of this interesting ecological niche can be found today.

The Midwest is a large tract of land that was, and is, the cradle of a vast agricultural and industrial complex. Unfortunately, nature has taken second place to progress. Visitors to this region will find outings limited in comparison to other regions. However, the few that are available are very precious.

Cleveland

Cleveland is home to thousands of corporations. It ranks third in the United States, behind only New York and Chicago. This brings thousands of visitors to the area on business.

Cleveland was built on a lake plain. Almost all the land along Lake Erie was once underneath the lake. The land is now a flat plain at the same elevation as

---

ABOVE: Visitors to the Cuyahoga Valley National Recreation Area near Cleveland walk by the Ritichie Ledges in Virginia Kendall Park. Photo courtesy of National Park Service.

the lake. Major drainage projects were necessary before cities could be built along the shoreline, and these projects destroyed most of the fragile shore ecology.

During the past quarter century, Cleveland has made a great effort to restore some natural settings in and around the city. It has more than 18,000 acres in its metropolitan park system, and it is recognized as a model city for its urban forestry program.

Cleveland's foresters have developed a process for selecting and planting trees that match the available air space. They have also developed a checklist of trees that should not be planted. While this approach will never see the region return to the majestic hardwood forests that once covered the land, it does provide enough trees to bring some bird life into the city's center.

Climate and Weather Information

The midwestern seasonal patterns are modified in Cleveland by the presence of Lake Erie. The winters are milder and the summers cooler than in most other midwestern cities of the same latitude.

Winters are cold and snowy, and spring is generally short. Summers are hot and humid. Fall presents the best weather for enjoying the outdoors, as the days are generally cool, but dry. Spring and summer are the wet seasons.

Getting Around Cleveland

Streets in Cleveland are laid out in a grid pattern, centered around Public Square. They are relatively easy to navigate. Downtown parking is limited, but there are a number of off-street parking facilities. Taxis are plentiful downtown, but the fares are more expensive than in other major cities.

The Cleveland Regional Transit Authority has bus or rapid transit train service to all parts of the city. Most of the outings listed that are in or just outside Cleveland can be reached by transit. For those further out you will need to rent a car. There are plenty of rental agencies in the city.

Indoor Activities

In addition to the activities listed, you may want to check colleges and universities in Cleveland and surrounding cities. Other arboretums and natural history and science museums may be of interest to you.

City of Cleveland Greenhouse
Rockefeller Park, 750 E. 88th St., Cleveland, OH 44108—216-664-3103

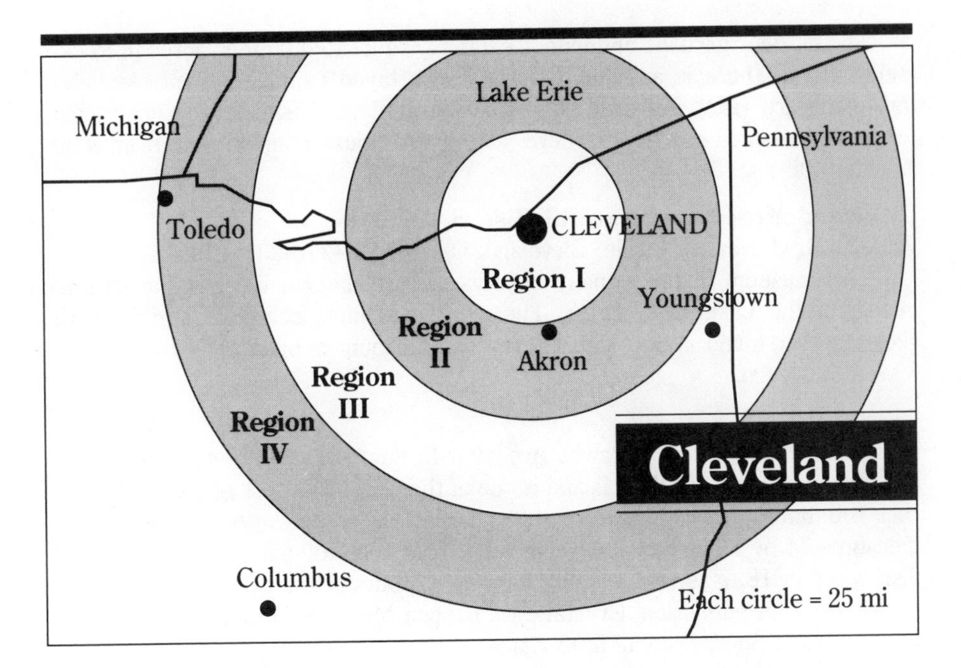

Indoor Activities

Region I
City of Cleveland Greenhouse
Cleveland Museum of Natural History
Garden Center of Greater Cleveland
Lake Erie Nature and Science Center

Region III
Crystal Cave

Outdoor Activities

Region I
Cascade Park
Cuyahoga Valley National Recreation Area
Gardenview Horticulture Park
Hinckley Reservation
Holden Arboretum
Metropark Huntington
Rocky River Reservation
Shaker Lakes Regional Nature Center
The Wilderness Center, Inc.

Region II
Nelson Kennedy Ledges State Park
Punderson State Park
West Branch State Park

Region III
East Harbor State Park
Galpin Wildlife and Bird Sanctuary
Kelleys Island State Park
Kingwood Center
Mill Creek Park
Mohican State Park
Seneca Caverns

Region IV
Crane Creek State Park
Maumee Bay State Park
Wildwood Preserve Metropark

The outdoor gardens and indoor exhibits are located on four acres in Rockefeller Park. There is a Talking Garden with taped explanations of plants so those who are vision-impaired can enjoy the gardens. Visitors are encouraged to touch and smell the flowers here. The green house is open daily from 9:30 A.M. to 4:00 P.M.

Cleveland Museum of Natural History
Wade Oval, University Circle, Cleveland, OH 44113—216-231-4600

This museum features many displays of early human life and the natural history of the Lake Erie region. The oldest and most complete human fossil skeleton ever found is on exhibit here. The museum is open daily from 10:00 A.M. to 5:00 P.M.

Crystal Cave
Heineman Winery Tour, Catawba Ave., Put-In-Bay, OH 43456—419-285-2811

This cave on South Bass Island is one of the world's largest geodes—a hollow rock with unusual crystalline formations inside. The largest crystal in this geode measures 24 by 18 inches and weighs approximately 300 pounds. The cave was discovered in 1897 as workers dug for water. Tours of the cave are included in the winery tour admission. Crystal Cave is open Monday–Saturday, 11:00 A.M.–5:00 P.M. and Sunday, noon–5:00 P.M.

Garden Center of Greater Cleveland
11030 East Blvd., Cleveland, OH 44106—216-721-1600

Plant exhibits, gardening lectures, and flower shows are featured inside the center and outside on the landscaped grounds. The gardens are open Monday–Friday, 9:00 A.M.–5:00 P.M. and Sunday, 2:00 P.M.–5:00 P.M.

Lake Erie Nature and Science Center
28728 Wolf Rd., Bay Village, OH 44140—216-871-2900

Located in Bay Village, just off State Route 6 and Interstate 90, this facility has live animal collections, displays, a petting zoo, and marine tank. Bird walks, animal shows, and nature hikes are offered on the weekends. A planetarium is also featured. The center is open Thursday–Tuesday, 1:00 P.M.–5:00 P.M.

Outdoor Activities

The activities listed cover a wider geographical area than most of the other cities in this guide. All of the Lake Erie shoreline in Ohio is included. Some of the outings can be reached on two-hour drives inland.

Cascade Park
c/o City Hall, 7303 Avon Belden Rd., Elyria, OH 44035—216-322-0926

This park has waterfalls, caves, interesting rock formations, and trails. These lead through wooded areas. Cascade Park is west of Cleveland off Interstate 90 on North Broad Street.

Cleveland Metropolitan Park System
4101 Fulton Pkwy., Cleveland, OH 44144—216-351-6300

This system has 12 parks connected over a 100-mile parkway, and contains almost 19,000 acres. Not all of these parks offer nature outings, but several do. Four interpretive centers have "Trails for All People" that accommodate the disabled. Contact the park headquarters for more information about the system.

Crane Creek State Park and Ottawa National Wildlife Refuge
14000 W. State Route 1, Oak Harbor, OH 43449—419-898-2495

This mostly sedge marsh is one of the top waterfowl observation spots in the Midwest. It is the last bald eagle nesting ground on the Great Lakes. The state park is 79 acres, and has hiking trails and picnicking. It is near Locust Point off State Route 2, east of Toledo, Ohio.

Cuyahoga Valley National Recreation Area
6699 Canal Rd., Valley View, OH 44125—216-524-1497

This recreation area contains 32,000 acres of river valley along a 22-mile section of the Cuyahoga River. Remnants of the Ohio and Erie Canal are in the area, which includes a floodplain with streams and creeks, forested valleys and upland plateaus. There is an interpretive center, plus hiking trails, canoeing, and two visitor centers. The Canal Visitor Center is located in the northwestern tip of the recreation area, and the Happy Days Visitor Center is on State Route 303. The visitor centers are open daily from 8:00 A.M. to 5:00 P.M. with longer hours in the summer. The recreation area is open daily, from dawn to dusk.

East Harbor State Park
1169 N. Buck Rd., Lakeside-Marblehead, OH 43440—419-734-4424

East Harbor State Park is seven miles east of Port Clinton off State Route 163. It sits along the shores of Lake Erie and features hiking trails and beach activities.

Galpin Wildlife and Bird Sanctuary
P.O. Box 424, Milan, OH 44846—419-499-2100

This sanctuary is about one-half mile southeast of Milan on Edison Drive. It features many varieties of trees, wildflowers, and birds. It is open daily.

Gardenview Horticulture Park
Strongsville, OH 44136—216-238-6653

South of Strongsville, on U.S. Route 42, this English-style garden shows uncommon and unusual plants. The park sits on 16 acres.

Hinckley Reservation
Hinckley, OH 44233—216-278-4181

Each year on March 15, the buzzards return to Hinckley at Whipp's Ledge. Then naturalists gather together for the largest organized bird walk in the country. This event began in 1958 after the *Cleveland Press* wrote a story about the punctual return of turkey vultures to the area. Hinckley town leaders proclaimed the Sunday after March 15th as Buzzard Sunday after thousands of tourists descended upon the town.

Holden Arboretum
Sperry Rd., Mentor, OH 44060—216-946-4400

This is a 3,100-acre natural museum with horticultural displays, gardens, woods, fields, lakes, and ravines. It has over 6,000 varieties of ornamental and native plants. There are nature and hiking trails that wind through the rolling hills and ravines. A visitor center and library are also featured. The arboretum is open Tuesday–Sunday, 10:00 A.M.–5:00 P.M.

Kelleys Island State Park
Kelleys Island, OH 43438—419-746-2546

Located off the shores of Lake Erie from Sandusky, Ohio, this is the largest island on the American side of Lake Erie. It includes a 659-acre park on the southern tip. The main features of the park are the glacial grooves left from the last Ice Age. They are considered among the best glacial carvings found in the United States. The glacial grooves have exposed fossilized marine life in the limestone bedrock of the island. During the spring and fall, daily ferry trips to the island from Sandusky are available. During the summer, the ferries leave from Marblehead.

Kingwood Center
900 Park Ave., West Mansfield, OH 44906—419-522-0211

Landscaped gardens, winding woodland, trails and two ponds with numerous waterfowl are featured at this 47-acre park. The gardens have one of the largest displays of tulips in the United States, and the greenhouses are filled with flowering plants. Kingwood Center is open daily from 8:00 A.M. to sundown.

Maumee Bay State Park
6505 Cedar Point Rd., Oregon, OH 44318—419-836-7758

Located eight miles northeast of Toledo off State Route 2, this shoreline park with wet woods and marshes is home to a wide variety of wildlife. The park features good hiking and bird watching.

Metropark Huntington
c/o Cleveland Metropolitan Park System, 4101 Fulton Pkwy., Cleveland, OH 44144—216-351-6300

Set on the shore of Lake Erie, this park has nature trails and the Lake Erie Nature and Science Center among its attractions. It is along U.S. Route 6 between Lake and Wolf Roads in Bay Village, Ohio.

Mill Creek Park
c/o Youngstown Department of Parks and Recreation, 26 S. Phelps St., Youngstown, OH 44503—216-744-4171

This large, 2,383-acre park on Glenwood Avenue has a six-mile gorge, three lakes, a waterfall, and a large formal garden. Numerous hiking trails lead to all areas of the park. The Ford Nature Center, also located in the park, has a semi-aquatic terrarium and aquariums with many native fish, plants, and crayfish. From March 1st to November 30th, the park is open Tuesday–Sunday, 10:00 A.M.–6:00 P.M. During the rest of the year, the park is open Saturday–Sunday, 10:00 A.M.–4:00 P.M.

Mohican State Park
P.O. Box 22, Loudonville, OH 44842—419-994-4290

This park is on State Route 97, five miles southwest of Loudonville. The Mohican River forms a large gorge, 1,000 feet wide at the top and 200 to 300 feet deep. Trails in the park lead to various parts of gorge.

Nelson Kennedy Ledges State Park
c/o Cuyahoga Valley National Recreation Area, 6699 Canal Rd., Valley View, OH 44125—216-524-1497

Located on State Route 282 near Garrettsville, this 167-acre park has unusual geologic formations and rare ferns. There is an outcropping of a long sandstone formation. The park is open daily year-round. From April 1st to October 31st, the hours are 6:00 A.M.–sunset. During the rest of the year, the hours are 8:00 A.M.–sunset.

Punderson State Park
P.O. Box 338, 10755 Kinsman Rd., Newbury, OH 44065—216-564-2279

Punderson State Park is ten miles south of Chardon on State Route 44. It covers 990 acres and has a 90-acre lake. Hiking trails lead to quiet portions of the park.

Rocky River Reservation
c/o Cleveland Metropolitan Park System, 4101 Fulton Pkwy., Cleveland, OH 44144—216-351-6300

This park follows the valley of the Rocky River south from Lake Erie. Part of the 100-mile park system that surrounds Cleveland, this park has a nature center and numerous trails.

Seneca Caverns
Bellevue, OH 44811—419-483-6711

Off State Route 269, five miles southwest of Bellevue, this is a large natural cavern that developed as a unique earth crack was created by geologic forces. The cavern has eight rooms on seven levels. It is open daily from Memorial Day to Labor Day and on weekends only during the rest of year.

Shaker Lakes Regional Nature Center
2600 S. Park Blvd., Shaker Heights, OH 44120—216-321-5935

This suburban center has nature and hiking trails, a bird observation area, and two human-made lakes on 300 acres. The center is open daily.

The Wilderness Center, Inc.
Wilmot, OH 44689—216-359-5235
The Wilderness Center, Inc. is one mile northwest of Wilmot on U.S. Route 250. This park contains an interpretive center with nature exhibits and lectures, and 570 acres of forests, prairies, and marshes. All of this is crisscrossed by trails which are open daily from dawn to dusk. The interpretive center is open Tuesday–Saturday, 9:00 A.M.–5:00 P.M., and Sunday, 1:00 P.M.–5:00 P.M.

West Branch State Park
5708 Esworthy Rd., Route 5, Ravenna, OH 44266—216-296-3239
This park is 12 miles east of Kent on State Route 5. It is a large, 8,002-acre, multi-use park with hiking trails along with a wide variety of other activities, such as fishing, boating, horseback riding, and swimming.

Wildwood Preserve Metropark
5100 W. Central Ave., Toledo, OH 43615—419-535-3058
This former estate is a 460-acre park which includes walking trails, nature programs, and a visitor center. The park is open daily from 7:00 A.M. to sunset.

Organizations That Lead Outings

Audubon Society of Greater Cleveland
140 Public Square, Cleveland, OH 44114—216-376-5550

Cleveland Waterfront Coalition, Inc.
401 Euclid Ave., Cleveland, OH 45219—216-771-2666

Sierra Club
13677 Old Pleasant Valley Rd., Cleveland, OH 44130—216-843-7272

A Special Outing

As stated in the introduction to this guide, you have to take nature where you find it. In Cleveland one place you can find it is the 285-acre Lake View Cemetery. There is no better geological outing in the region than this cemetery, and the bird watching is as good as any place in or near Cleveland.

During the late Denovian Period, a shallow sea once covered the region. Offshore banks from this sea may have deposited sandstone here. This sandstone and outcroppings of the Bedford Formation can be seen in several areas throughout the cenetery. Ripple marks left by waves in the shallow water are still visible in some of the exposed layers.

The steep cliffs of the surrounding hills expose parts of the Chagrin Formation. Cleveland shale can be seen at the site of the dam that was built on Dugaway Brook in the 1970s.

In addition to the native geological formations, the tombstones in the cemetery are representative of all three major rock types. Lake View Cemetary is located at 12316 Euclid Avenue, Cleveland, Ohio 45236, 216-421-2665.

Nature Information

Cleveland Convention and Visitors Bureau
1301 E. 6th St., Cleveland, OH 44114—216-621-4110

Division of Parks and Recreation
Department of Natural Resources, Fountain Square, Bldg. C, Columbus, OH 43224—614-265-7000

Division of Wildlife
Department of Natural Resources, Fountain Square, Bldg. C, Columbus, OH 43224—614-265-6300

Manager, Lake Department
Muskingum Conservancy District, New Philadelphia, OH 44663—216-343-6647

Ohio Department of Development
Division of Travel and Tourism, P.O. Box 1001, Columbus, OH 43266-0101—800-BUCKEYE

Further Reading

Cantor, George. *The Great Lakes Guidebook: Lakes Ontario and Erie.* Ann Arbor: University of Michigan Press, 1978.

Hough, Jack L. *Geology of the Great Lakes.* Champaign: University of Illinois Press, 1966.

McKee, Russell. *Great Lakes Country.* New York: Crowell, 1966.

McCaig, Barbara, and Margie McCaig. *Ohio State Parks and Forests.* Wauwatosa, WI: Affordable Adventures, Inc., 1987.

Zimmerman, George. *Ohio: Off the Beaten Path.* Chester, CT: The Globe Pequot Press, 1987.

Chicago

Chicago has long been known as "The Second City," and has often chafed under that label. Among the cities in this guide, it cannot even claim second city status in nature activities available to residents and visitors.

Chicago's earliest motto meant "City in A Garden," and it attempted to live up to this motto with its extensive park system. While many of the grand plans for parks and gardens in the city have fallen by the wayside, Chicago still boasts of one of the outstanding park systems in the nation. This system is devoted to recreation, however, and offers little to those who want to have a naturalist-type outing.

As the city grew, the natural marshes and swamps that once bordered Lake Michigan were filled. The growth of Chicago eventually engulfed the prairie that surrounded those marshes. The lakefront is completely developed for miles on

ABOVE: The Illinois and Michigan Canal offers visitors a chance to experience history as well as nature. Sections such as this one near Lock 6 offer excellent canoeing in addition to canalside trails. Photo by Illinois Department of Conservation.

either side of Chicago, and the forests that once covered the rolling, glacial carved land have been removed to make room for productive farmland.

On the plus side, Chicago is known for its many fine museums, and natural history is well represented. Its gardens and conservatories are also some of the finest in the United States.

Climate and Weather Information

Unpredictable is the word for Chicago's weather. While summers are usually hot and humid, it is not unusual to spend several weeks in mid-summer without seeing the thermometer climb above 75°F. Winters generally see plenty of snow. Moderate winds off the lake often drop the wind chill factor below zero.

Spring and fall are the most pleasant seasons in the city, but even these have an unpredictable quality. Spring can offer warm days when the crocus begin to bloom, and then winter makes a dramatic return, even as late as May. Fall may last into late November, or may come to an abrupt halt by mid-October.

Getting Around Chicago

Chicago's transit system is one of the best in the United States. It doesn't extend far enough away from the city for it to be beneficial to visitors who want to have a naturalist's outing—unless such an outing is to one of the many fine museums and conservatories in the city itself.

Driving downtown can be frustrating unless you know exactly where you want to go. Rush hour, as in other large cities, presents some of the worst traffic. Street parking is almost nonexistent, and parking lots, both public and private, charge above average rates. Taxi fares can be expensive in Chicago, although cabs are numerous. Rental car agencies are plentiful, but you may want to rent on the outskirts of the city to avoid parking and driving problems.

Indoor Activities

Chicago is known for its museums, and many of these offer nature exhibits. The following list includes interesting places you can visit during the peak of summer or winter.

Adler Planetarium
1300 S. Lake Shore Dr., Chicago, IL 60605—312-322-0300

This museum has both historical and modern collections. A solar telescope can be used by visitors during the summer to view solar flares and sunspots. A multimedia sky show is presented daily. The planetarium is open daily from 10:00 A.M. to 5:00 P.M.

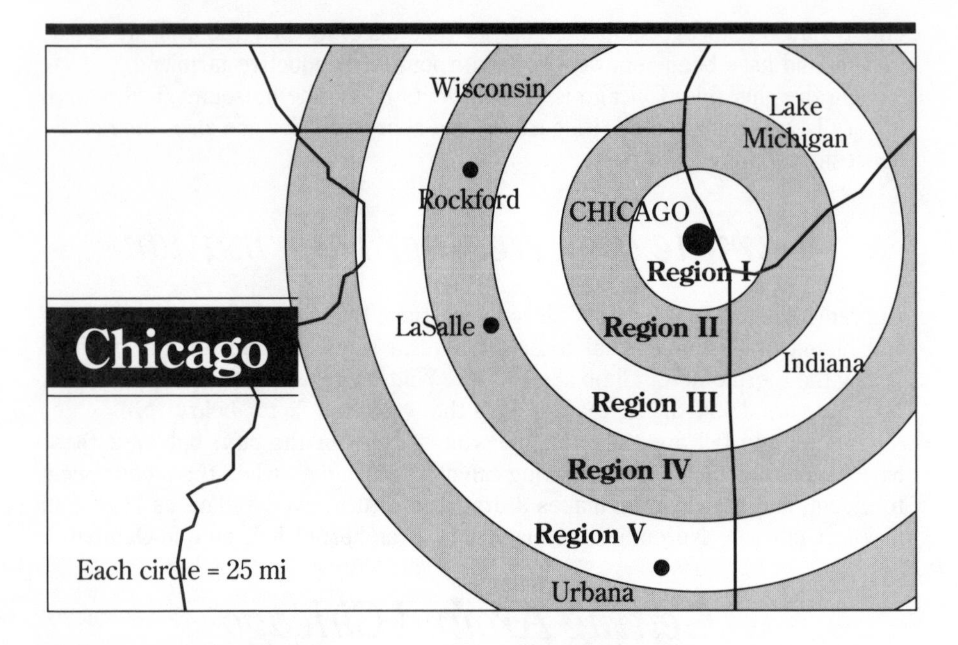

Indoor Activities

Region I

Adler Planetarium
Chicago Academy of Science
Dearborn Observatory
Evanston Ecology Center
Field Museum of Natural History
Garfield Park Conservatory
Hillary S. Jurica Memorial Biology Museum
John G. Shedd Aquarium
Lincoln Park Conservatory
Oak Park Conservatory
Trailside Museum of Natural History

Region IV

Burpee Museum of Natural History

Outdoor Activities

Region I

Chicago Botanic Gardens
Illinois Prairie Path
Ladd Arboretum
Lincoln Park
Morton Aquarium
Willowbrook Wildlife Haven
Wooded Island

Region II

Burnidge Forest Preserve
Des Plaines Fish and Wildlife Area
Fermi National Accelerator Laboratory
Illinois Beach State Park
Moraine Hills State Park
Silver Springs Fish and Wildlife Area
Tyler Creek Forest Preserve
Volo Bog State Natural Area

Region III

Chain O'Lakes State Park
Gebhard Woods State Park
Goose Lake Prairie Natural Area
Illini State Park
Indiana Dunes National Lakeshore
Indiana Dunes State Park
Kankakee River State Park
Rock Cut State Park

Region IV

Matthiessen State Park Nature Area
Starved Rock State Park

Region V

Forest Glen Preserve
Kickapoo State Park
Marshall State Fish and Wildlife Area

Burpee Museum of Natural History
813 N. Main St., Rockford, IL 61103—815-965-3132

About 90 miles west of Chicago, this museum's main collections are related to Indians, animals, fossils, and rocks. It is open Tuesday–Saturday, 1:00 P.M.–5:00 P.M. and Sunday, 2:00 P.M.–5:00 P.M.

Chicago Academy of Science
2001 N. Clark St., Chicago, IL 60614—312-549-0606

This museum specializes in the wildlife of the Great Lakes region including prehistoric animals. It is open daily from 10:00 A.M. to 5:00 P.M.

Dearborn Observatory
2131 Sheridan Rd., Evanston, IL 60201—708-492-3173

This observatory includes a Civil War-era 18 and one-half-inch refracting telescope. It is in a corner of Northwestern University's campus. With prior reservations, visitors can view the skies on clear Friday evenings from April to October. Phone for hours.

Evanston Ecology Center
2024 McCormick Blvd., Evanston, IL 60201—708-864-5181

Slide shows and self-guiding tours at this facility demonstrate ecologically sound living techniques. The Energy Information Center provides booklets and information about solar and wind energy. It is located in Ladd Arboretum. See Outdoor Activities section for further details. Open daily from 9:00 A.M. to 5:00 P.M.

Field Museum of Natural History
Roosevelt Rd. & Lake Shore Dr., Chicago, IL 60605—312-922-9410

More than ten acres of exhibits covering most of the Earth's history are on display in this museum, which opened in 1893. It ranks among the foremost natural history museums in the world. The Field Museum is open daily from 9:00 A.M. to 5:00 P.M.

Garfield Park Conservatory
300 N. Central Park Ave., Chicago, IL 60624—312-533-1281

This is the largest conservatory in the world under one roof, with over four and one-half acres. Over 5,000 varieties of plants are kept here. There are eight exhibit halls. There is also a special garden for the blind. The conservatory is open daily from 9:00 A.M. to 5:00 P.M.

Hillary S. Jurica Memorial Biology Museum
Science Center, Illinois Benedictine College, 5700 College Rd., Lisle, IL 60532—708-968-7270

This museum exhibits many bird, insect, and mammals specimens, as well as minerals and fossils. The specimens were collected by the late biologist and educator Hillary S. Jurica. Hours are by appointment.

John G. Shedd Aquarium
1200 S. Lake Shore Dr., Chicago, IL 60605—312-939-2438

More than 8,000 freshwater and marine animals in over 200 settings are on display here. There is a 90,000-gallon Coral Reef Exhibit where divers hand-feed the fish. Shedd Aquarium is open daily from 10:00 A.M. to 5:00 P.M.

Lincoln Park Conservatory
2400 N. Stockton Dr., Chicago, IL 60614—312-968-4770

This is another of the excellent conservatories which is part of the Chicago Park system. This conservatory is smaller than the one in Garfield Park, but has some ancient specimens in its five halls. It is open daily from 9:00 A.M. to 5:00 P.M.

Oak Park Conservatory
621 Garfield St., Oak Park, IL 60304—708-386-4700

Once threatened with the wrecking ball, this conservatory was saved by preservationists. It now has a central tropical house with a fish pond, a fern house with temperate zone plants, and a desert house with a smell-and-touch herb garden. The conservatory is open daily from 9:00 A.M. to 5:00 P.M.

Trailside Museum of Natural History
738 Thatcher Ave., River Forest, IL 60305—708-366-6530

The museum has exhibits on local flora and fauna, as well as interpretive programs. It is open from 10:00 A.M. to 4:00 P.M. daily, except Thursday.

Outdoor Activities

When you participate in outdoor nature activities around Chicago you generally must travel away from the central part of the city. The farther away you travel, the better your chances are of finding an outing that satisfies your needs.

Burnidge Forest Preserve
Coombs Road, Elgin, IL 60123—708-741-4833

This 337-acre preserve has hiking trails for observing a variety of birds and wildlife. The trails are used for cross-country skiing, snowmobiling, and tobogganing in the winter.

Chain O'Lakes State Park
39947 N. State Park Rd., Spring Grove, IL 60081—708-587-5512

This park is off State Route 173, about 20 miles west of Lake Michigan and four miles south of the Illinois–Wisconsin border. The largest concentration of natural lakes in Illinois is found in this 2,793-acre park and 3,230-acre conservation area. The ten lakes in the region have a combined acreage of 6,465, and almost 500 miles of shoreline.

Within the park and conservation area, the land is primarily deep peat deposits covered by a freshwater bog. Higher land in the park was shaped by glaciers, and slopes up to 200 feet above the bogs. These uplands have mixed hardwood and pine forests that are home to numerous species of wildlife. There are over two miles of nature trails within the park, and plenty of opportunity to hike along the Fox River.

Chicago Botanic Gardens
Lake-Cook Rd., Glencoe, IL 60022—708-835-5440

With over 300 acres of hills, lakes, and islands, these landscaped gardens have many attractions. There are demonstrations gardens, a nature trail, a prairie area, and an education center where classes, lectures, and special events are held. The gardens are open daily from 8:00 A.M. to sunset.

Des Plaines Fish and Wildlife Area
R.R. 3, P.O. Box 167, Wilmington, IL 60481—815-243-5326

About ten miles south of Joliet off Interstate 55 this is primarily a fishing and hunting site. There are also several natural areas set aside in the 4,900 acres of the area. Be careful during the hunting season.

Fermi National Accelerator Laboratory
Kirk Rd. & Pine St., Batavia, IL 60510—708-840-3000

Fermilab is one of the largest particle accelerators in the world, but it also has 640 acres of restored prairie within its circular enclosure. Swans, ducks, and buffalo roam the grounds. Visitors are welcome daily from 8:00 A.M. to 5:00 P.M.

Forest Glen Preserve
R.R. 1, P.O. Box 495A, Westville, IL 61883—217-662-2142

About ten miles southeast of Danville, Illinois, this nature preserve is located on the Vermillion River. It includes camping, trails, a nature center, and an observation tower that overlooks the river.

Gebhard Woods State Park
P.O. Box 272, Morris, IL 60450—815-942-0796

Off State Route 47 on the north bank of the Illinois and Michigan Canal, this 30-acre park is part of the I&M Canal State Trail. The park offers fishing, bird watching, and picnicking. The largest tree in Illinois, a 120-foot cottonwood that has a circumference of 27 feet 4 inches, is one mile west of the park on the south side of the canal.

Goose Lake Prairie Natural Area
50110 N. Jugtown, Morris, IL 60450—815-942-2899

South of the Illinois River and eight miles southeast of Morris, this is one of the last remains of prairie in Illinois. It is one of the largest preserves of tall grass prairie in the nation. Of the park's 2,357 acres, 1,513 are dedicated as an Illinois Nature Preserve. Visitors to this park will not find a lake, since it was drained at the end of the last century for farming, but they will find grasses and flowers much like those the earliest settlers in the region found. A tall grass prairie has wildflowers almost year-round, and each season has its own special hue.

Illini State Park

R.R. 1, P.O. Box 60, Marseilles, IL 61341—815-795-2448

Near State Route 6 and across the Illinois River from Marseilles in La Salle County, this 510-acre park is a popular recreation site not far from Starved Rock State Park. Its hardwood forests are home to a wide variety of wildlife. A half-mile interpretive trail, and a two-mile hiking trail lead visitors through the riverside forests.

Illinois Beach State Park

Zion, IL 60099—708-662-4811

The first Illinois nature preserve was established here in 1964. Today the two units of the park have 4,160 acres that stretch six and one-half miles along the shores of Lake Michigan north of Waukegan, Illinois.

An interpretive center is at the Nature Area, which is located in the southern unit of the park. Nature walks and talks are provided there year-round. The park provides some of the best bird watching in the Chicago area, and the marshlands and the Dead River are home to numerous species of fish.

Illinois Prairie Path

Elmhurst, IL 60126—708-665-5310

Beginning in Elmhurst, Illinois, along York Road, this 45-mile long biking and hiking trail follows an old railroad right-of-way. Wildflowers, prairie grasses, and birds are plentiful. Call for maps and information about the path.

Indiana Dunes National Lakeshore

1100 N. Mineral Springs Rd., Porter, IN 46304—219-926-7561

This 12,000-acre park extends along ten miles of Lake Michigan's southern shore. The park protects the last undeveloped area of Indiana sand dunes that were formed some 15,000 years ago during the last Ice Age. Visitors can enjoy a nature preserve with visitor center, excellent bird watching, and many hiking trails that lead through lakeside bogs, marshes, and sand dunes.

Indiana Dunes State Park

1600 N and 25 E, Chesterton, IN 46304—219-926-4520

This 2,182-acre park includes 192-foot Mt. Tom, the highest remaining Indiana sand dune. Two-thirds of the park was declared a state nature preserve in 1976, thereby guaranteeing that it would remain undeveloped. Over eight miles of marked hiking trails and a nature center are popular features here.

Kankakee River State Park

R.R. 1, Bourbonnais, IL 60914—815-933-1383

Located six miles northwest of Kankakee on State Routes 102 and 113, this 2,780-acre park stretches over 11 miles of the Kankakee River. Waterfalls with limestone outcroppings and colorful rare plants dot the landscape. These and prairie glades are only a few of the natural wonders that visitors to this park enjoy. The visitors center provides guided nature walks and talks.

Kickapoo State Park

R.R. 1, P.O. Box 374, Oakwood, IL 61858—217-442-4915

About four miles west of Danville on U.S. 150, this 2,843-acre park features multiple activities including hiking and nature trails. A naturalist is also on staff.

Ladd Arboretum

2024 McCormick Blvd., Evanston, IL 60201—708-864-5181

The arboretum's 23 acres along the North Shore Channel are divided between a bird sanctuary and formal gardens. Hiking paths lead through both sections. The Evanston Ecology Center (see Indoor Activities) is located on the grounds. The arboretum is open daily from 9:00 A.M. to 5:00 P.M.

Lincoln Park

2400 N. Stockton Dr., Chicago, IL 60614—312-294-4770

The largest of Chicago's parks, Lincoln Park has something for everyone. For the naturalist, there is the Zoo Rookery, a bird sanctuary just to the north of the Lincoln Park Zoo along Cannon Drive.

Marshall State Fish and Wildlife Area

R.R. 1, P.O. Box 238, Lacon, IL 61540—309-246-8351

This 6,000-acre area is primarily for hunting and fishing, but there are a number of naturalist sites along the bottomlands and bluffs of this stretch of the Illinois River. Numerous birds and animals live along the river. Cottonwood and silver maple are the main trees in this area. There is a three-and-one-half-mile hiking trail through several habitats. In season, this trail is heavily used by hunters.

Matthiessen State Park Nature Area

P.O. Box 381, Utica, IL 61373—815-667-4868

Off State Route 178, seven miles southeast La Salle, this 1,629-acre nature site has waterfalls, cliffs, and canyon trails. Bird watching and wildflowers walks are favorite activities. The park is open daily from 6:00 A.M. to 10:00 P.M.

Moraine Hills State Park

914 S. River Rd., McHenry, IL 60050—815-385-1624

This park sits alongside the Fox River in the Chain O'Lakes region of Illinois and contains many natural features. An interpretive center is located in the park office. Over 11 miles of hiking trails lead through the park. The half-mile Pike Marsh Nature Trail, with its 1,300-foot boardwalk, describes the features of Lake Defiance—one of the few undisturbed glacial lakes in Illinois. The 120-acre Leatherleaf Bog is an excellent example of a kettle moraine. Several dedicated natural areas, including a prairie site, offer excellent opportunities to view relatively undisturbed examples of Illinois habitats.

Morton Arboretum

State Route 53, Lisle, IL 60532—708-968-0074

This 1,500-acre arboretum includes about eight miles of roads and 13 miles of trails that give visitors access to the wildlife sanctuaries, cultivated gardens, and native prairie. There is an orientation slide show, as well as a trail guide, at the visitors center. Morton Arboretum is open daily from 9:00 A.M. to 7:00 P.M.

Rock Cut State Park

7318 Harlem Rd., Caledonia, IL 61011—815-885-3311

This 2,743-acre park is home to abundant animal and plant life. Hikers can see beavers, deer, foxes, and raccoons, as well as a wide variety of wildflowers in the hardwood forests. There is also a large human-made lake in the park. A number of nature, interpretive, and hiking trails traverse the area.

Silver Springs Fish and Wildlife Area

R.R. 1, P.O. Box 318, Yorkville, IL 60560—708-553-6297

Five miles west of Yorkville off State Route 47, and south of U.S. Route 30, this multi-purpose unit not only offers hunting and fishing, but it also has peaceful nature trails. It is home to numerous migratory waterfowl, herons, songbirds, and deer. One section of the area is being restored to native prairie vegetation, and a wildlife viewing area has been developed near Loon Lake. There are three and one-half miles of hiking trails, and a seven-mile horseback riding trail that can also be used for hiking. During the hunting season, hikers should be cautious on both trails.

Starved Rock State Park

P.O. Box 116, Utica, IL 61373—815-667-4726

Off State Route 71, five miles east of La Salle, this is one of the most popular state parks in Illinois. The park gets its name from a 125-foot high rock where American explorer Sieur de La Salle built Fort St. Louis in 1682. In 1769 a band of Indians took refuge on the summit of the rock, where they died from lack of food and water.

The 2,630 acres in the park are crisscrossed by over 20 miles of hiking trails. These trails lead into the 18 canyons in the park that were formed by streams pouring into the Illinois River. Waterfalls form at the head of each canyon each spring, and often reappear in the summer after heavy rains. Hiking is not permitted in unmarked areas of the park.

There are many unique rock formations in the park, most of which are composed of St. Peter's sandstone. A massive upfold brought this layer of sandstone to the surface about 700,000 years ago. Glaciers have since leveled the surrounding land.

The wide variety of shelters and food in the park nurture to many native plants and animals. Other state parks in La Salle County include Buffalo Rock State Park, Illini State Park, and Matthiessen State Park.

Tyler Creek Forest Preserve
Elgin, IL 60120—708-741-5082

Off State Route 31 near Elgin, Illinois, this is a 50-acre preserve with hiking trails. It sits on a bluff, and has a variety of native nut trees.

Volo Bog State Natural Area
28478 Brandenburg Rd., Ingleside, IL 60041—815-344-1294

A quaking bog is the feature attraction at this 869-acre wilderness area. The main bog is about 50 acres. Walking directly on the spongy vegetation of the bog is like walking on a waterbed. Visitors are required to stay on the half-mile trail that loops around the bog. A boardwalk, resting on Styrofoam pontoons, crosses the wetter portions of the bog. Many wildflowers are seen along the trail, and these change by varying habitat in the area. Cranes, herons, and blackbirds are among over 200 species of birds in the area.

A summer program includes guided bird and wildflower tours, and a bat walk to an old barn. The trail is open year-round. The visitor center is open Wednesday–Sunday from Memorial Day to Labor Day and on weekends and holidays during the rest of the year. Volo Bog State Natural Area is off U.S. Route 12 and State Route 120 north of Chicago.

Willowbrook Wildlife Haven
525 S. Park Blvd., Glen Ellyn, IL 60137—708-790-4900

Wounded, lost, and captured wildlife are cared for at this facility. Visitors enjoy guided tours and talks. Willowbrook Wildlife Haven is open daily from 9:00 A.M. to 5:00 P.M.

Wooded Island
57th St. & Lake Shore Dr., Chicago, IL 60605—312-493-7058

The Clarence Darrow Memorial Bridge connects this island to the mainland. The island, south of the Museum of Science and Industry, is one of the few natural areas of Chicago. Flora and fauna abound on the island, and early morning bird-watching walks are led from spring through fall. Call on weekdays between 5:00 P.M. and 6:00 P.M. pm for information on these.

Organizations That Lead Outings

American Youth Hostels
3712 N. Clark St., Chicago, IL 60657—312-327-8114

Chicago Audubon Society
5801 N. Pulaski Rd., Chicago, IL 60646—312-539-6793

Field Museum of Natural History
Kroc Environmental Field Trips, Roosevelt Rd. & Lake Shore Dr., Chicago, IL 60605—312-922-9410, Ext. 363

Nature Conservancy
79 W. Monroe, Chicago, IL 60603—312-236-7523

Reed's Canoe Trips, 907 N. Indiana Ave., Kankakee, IL 60901—815-932-2663

Sierra Club
506 S. Wabash, Chicago, IL 60605—312-431-0158

A Special Outing

The Illinois and Michigan Canal was once the principal link between the Chicago and Illinois Rivers. It was the final link of a continuous waterway from the Atlantic Ocean to the Gulf of Mexico. Between 1836 and 1848, the canal was dug completely by hand. Its 96 miles traversed the lowland divide between Lake Michigan and the rivers of Illinois that flowed south into the Mississippi. It provided a means of traveling from the Atlantic through the Erie Canal, across the Great Lakes, and down the Mississippi to the Gulf of Mexico.

In 1974 management of the canal was transferred to the Illinois Department of Conservation for the development of a 61-mile hiking and biking trail, and a 28-mile canal that is perfect for calm water canoeing. Now known as the Illinois and Michigan Canal National Heritage Corridor, the route runs from Chicago to Peru, Illinois. It features historic sites, residential neighborhoods, forest preserves, and 39 natural areas of rare flora and fauna. Among these is the Lockport Prairie, where some of the best examples of native prairie grasses and wildflowers are found.

A number of recreational areas are located along the corridor, and these offer excellent sites for beginning a nature outing. One of these is Gebhard Woods, where the largest tree in Illinois is found. Among many other imposing trees in the hardwood forests, blooms of bluebell, white trout lily, trillium, violets, wild ginger, and spring beauty carpet the grounds during the spring. A wide variety of birds and small mammals use the forests as home.

The William G. Stratton unit offers an excellent area for launching canoes and small boats, and is noted for its river fishing. Buffalo Rock State Park is set on a bluff that was once an island in the Illinois River. It now offers a splendid view of the river below. At the peak of Buffalo Rock, the sandstone bluff is 500 to 1,000 feet wide, and is covered with a cedar, oak, and pine forest.

Channahon is the site of two locks of the old canal. Fishing is available in both the canal and the nearby DuPage River. A 15-mile stretch of the towpath is readily accessible for further exploring.

All in all, the trails along the old canal offer visitors to Chicago the easiest access to a cross section of Illinois wilderness activities. Opportunities to view areas that contain fine examples of the state's prairie flora before the rich land was turned into large tracts of agricultural fields abound. For more information write the Illinois and Michigan Canal, P.O. Box 272, Morris, IL 60450; or call 815-942-6529.

Nature Information

Chicago Park District
425 E. McFetridge Dr., Chicago, IL 60605—312-294-2270

Chicago Tourism Council
163 E. Pearson Ave., Chicago, IL 60611—312-280-5740

Forest Preserve District
536 N. Harlem Ave., River Forest, IL 60305—708-261-8400

Illinois Department of Conservation
524 S. Second St., Springfield, IL 62706—217-782-6302

Illinois Tourist Information Center
310 S. Michigan Ave., Chicago, IL 60604—312-793-2094

Further Reading

Fensom, Rod, and Julie Foreman. *Illinois: Off the Beaten Path.* Chester, CT: The Globe Pequot Press, 1987.

Fodor's: Chicago and the Great Lakes Recreation Areas. New York: David McKay, 1983.

Mark, Norman. *Norman Mark's Chicago: Walking, Bicycling, and Driving Tours of the City.* Chicago: Chicago Review Press, 1987.

Schaeffer, Norma, and Kay Franklin. *'Round and About the Dunes.* Beverly Shores, IN: Dunes Enterprises, 1984.

Soli, Lynn, Barb McCaig, and Margie McCaig. *Illinois Parks and Forests.* Wauwatosa, WI: Affordable Adventures, Inc., 1986.

Thomas, Bill, and Phyllis Thomas. *Natural Chicago.* New York: Harper and Row, 1987.

Minneapolis-St. Paul

Thomas Wirth, the first superintendent of Parks and Recreation for Minneapolis, was a visionary who wanted to protect and develop a greenway along both sides of Minnehaha Creek through the heart of Minneapolis. While he did not get everything he wanted, he did get the basis of a park system that is now one of the best in the United States. It is possible for visitors to Minneapolis–St. Paul to experience the feeling of the great North Woods without having to stray far from the thriving urban centers of these two cities, separated by the Mississippi River.

Minnesota's land surface has been gouged, planed, and rearranged by glaciers at least four different times in recent geological history. The evidence of the advances and retreats of these glaciers can be seen in the rocky bluffs, fertile

ABOVE: Visitors can take contemplative walks along the water during the early morning in the Blacklock area of the Minnesota Valley National Wildlife Refuge. Photo courtesy of U.S. Fish and Wildlife Service.

valleys, and deep-soiled prairies of the state. But nowhere is this evidence more apparent than in and around more than 15,000 lakes spread throughout northern Minnesota.

The North Woods are famous for their dense evergreen forests and abundant lakes. While Minneapolis and St. Paul sit on the southern edge of this region, visitors have many opportunities to explore lakes, rivers, and forests both in the cities, and in nearby reserves and parks.

Nearly 1,000 lakes and 500 parks are located within a short distance of the Minneapolis–St. Paul urban area. These have miles of hiking and nature trails that lead naturalists into pristine settings where the native flora and fauna of Minnesota can be viewed.

Both cities have maintained or restored river and lake waterfronts so their natural beauty can be enjoyed. Residents and visitors share the many campgrounds, nature trails, and other outdoor activities that abound in the parks of the surrounding counties. Lake Minnetonka, west of the Twin Cities, and White Bear Lake, to the north, were once resort centers that attracted visitors from as far away as New Orleans. They still draw many visitors who want to explore nature.

Inside the cities there are almost 250 parks; Minneapolis has an acre of park for every 80 inhabitants. There are 30 lakes within a 30-minute drive of downtown St. Paul.

Climate and Weather Information

Cold is the dominant thought that comes to mind when people think about the weather in Minneapolis–St. Paul. While it is true that the winters are long and harsh, visitors can enjoy the outdoors much of the year.

Summer days are often hot and rather humid, but evenings are generally cool and pleasant. The heat and humidity, along with the long hours of daylight, bring dense swarms of flying and biting insects to the lakes, rivers, and forests. Visitors should bring adequate repellent and clothing to combat these.

Both spring and fall are short seasons. Chilly days can last as late as the end of April, and come as early as October. The long winter has generally settled in by November, snow begins to fall heavily by December, and temperatures from December to March average between 2° and 25°F. There are many days in January and February when temperatures are far below zero.

Because Minnesota residents are used to the long cold winters, they have developed their park system so that it offers opportunities to participate in outdoor activities, even in sub-zero weather. Few other northern cities provide similar opportunities. You should take advantage of them in the Minneapolis–St. Paul region.

Getting Around Minneapolis–St. Paul

Both cities are relatively easy to drive in, expecially when compared with other cities in this guide. A good street map, however, is advisable. Taxis can only be ordered by telephone or hired at taxi stands at reasonable rates.

The Metropolitan Transit System operates throughout the area. Many of the parks and lakes in and near the Twin Cities are accessible to visitors who do not wish to rent cars or to travel outside the cities. There are numerous car rental agencies for those who wish to travel farther away from downtown. The relative ease of driving in the cities makes it possible to rent a car near downtown and not have to worry about traffic conditions.

Indoor Activities

Many visitors are reluctant to venture out during winter in the Twin Cities. They welcome a chance to visit some indoor facilities.

James Ford Bell Museum of Natural History
17th Ave. & University Ave., Southeast, Minneapolis, MN 55455—612-624-1852

This museum exhibits Minnesota plants and animals. Major wildlife art and photography exhibits, as well as ones on natural history research, change frequently. It is open Tuesday–Saturday, 9:00 A.M.–5:00 P.M. and Sunday, 1:00 P.M.–5:00 P.M.

Minneapolis Planetarium
300 Nicollet Mall, Minneapolis, MN 55401—612-372-6644

Each show includes a 15-minute review of stars in the Minneapolis sky that evening. Presentations of major discoveries and space explorations are also featured. Programs run Monday–Friday at 11:00 A.M. and 2:00 P.M., Thursday at 7:30 P.M., and Saturday–Sunday at 2:00 P.M.

Minnesota Zoo
12101 Johnny Cake Rd., Apple Valley, MN 55124—612-432-9000

The Minnesota Trail, one of six separate trails in this 480-acre zoo, is an excellent introduction to native animals. During the winter visitors can cross-country ski along the walkways among the exhibits. The zoo is open daily from 10:00 A.M. to 6:00 P.M.

Science Museum of Minnesota
30 E. 10th St., St. Paul, MN 55101—612-221-9400

This museum has displays on anthropology, biology, and the natural sciences, in addition to technology and physical sciences. It is open Monday–Saturday, 9:30 A.M.–9:00 P.M. and Sunday, 11:00 A.M.–9:00 P.M.

Town Square Park
7th St. & Minnesota Blvd., St. Paul, MN 56304—612-292-7400

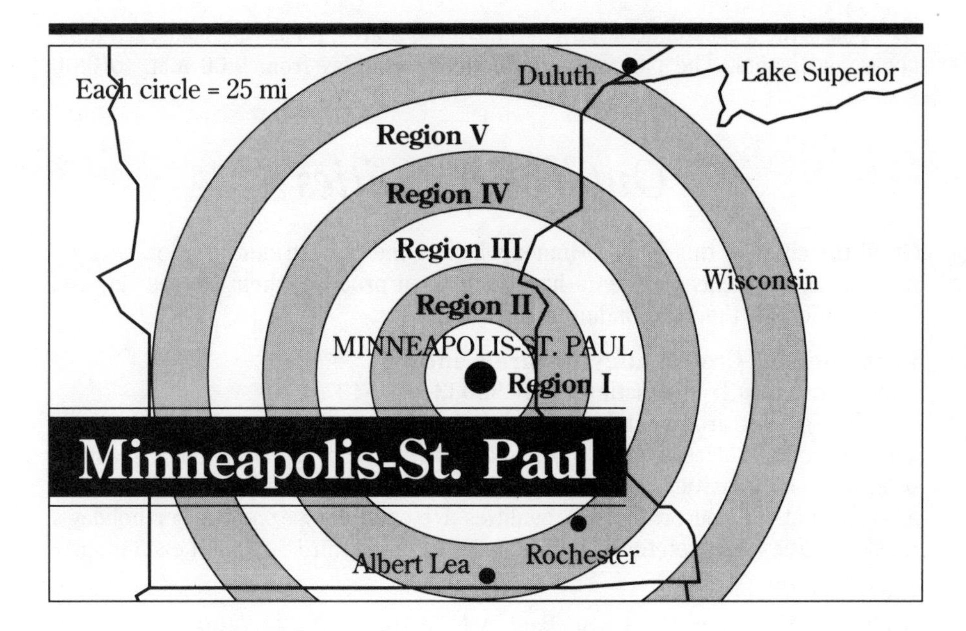

Minneapolis-St. Paul

Indoor Activities

Region I

James Ford Bell Museum of Natural History
Minneapolis Planetarium
Minnesota Zoo
Science Museum of Minnesota
Town Square Park

Outdoor Activities

Region I

Como Park
Eloise Butler Wildlife Garden and Bird Sanctuary
Fort Snelling State Park
Lakes Cedar, Harriet, Calhoun, and Lake of the Isles
Lakes Hyland, Bush, and Bryant
Lilydale–Harriet Island Regional Park
Lyndale Park
Minnehaha Park
Nokomis–Hiawatha Regional Park
Sibley Park
St. Anthony Falls
Theodore Wirth Park
University of Minnesota Landscape Arboretum
Upper Sioux Agency State Park

Region II

Carpenter St. Croix Valley Nature Center
Dalles of the St. Croix
Frontenac State Park
Interstate State Park
Nerstrand Woods State Park
River Bend Nature Center
St. Croix and Lower St. Croix National Scenic Riverway
St. Croix State Park
Wild River State Park
William O'Brien State Park

Region III

Gopher Campfire Game Sanctuary
Heritage Park
Lake Pepin
Minneopa State Park
Sherburne National Wildlife Refuge
Sibley State Park

Region IV

Fountain Lake
Helmer Myre State Park
J.C. Hormel Nature Center
Lac qui Parie State Park
Mille Lacs Kathio State Park
Pine Grove Park

Region V

Rice Lake National Wildlife Refuge
Savanna Portage State Park

This is the largest indoor park in the United States. It has waterfalls, streams, pools, and plants. The park is open Monday–Saturday from 9:00 A.M. to 9:00 P.M.

Outdoor Activities

Of all the cities in this guide, Minneapolis has the closest identification with its natural environment. Residents have long been proud of their interest in, and preservation of, the surrounding country.

Carpenter St. Croix Valley Nature Center
12805 St. Croix Trail, Hastings, MN 55033—612-437-4359

This is an environmental education center with more than 15 miles of trails and one mile of access to the St. Croix River. Bird banding, many seasonal programs and activities, plus rehabilitation of injured and captured raptors are all offered at the center. Outside facilities are open daily except major holidays. Inside facilities are open to the public on the first and third Sunday of each month.

Como Park
Snelling & Lexington Pkwy., St. Paul, MN 55104—612-292-7400

This 448-acre park includes a 70-acre lake, a conservatory that is open year-round, and a zoo. One of the six trails at the zoo is the Minnesota Trail. It features native Minnesota animals such as beavers, otters, lynx, weasels, and many more medium-to-small animals from nearby lake, prairie, and forest habitats.

Dalles of the St. Croix
Taylor Falls, MN 55084—612-465-5711

Unusual rock formations form the St. Croix River gorge. The shale and limestone formations contain some early fossil layers with trilobites, pteropods, and cystoids. At the Dalles, lava cliffs rise sharply above the river for over 200 feet. The river reaches a depth of about 100 feet here. The gorge is off U.S. 8 near the Minnesota-Wisconsin border.

Fort Snelling State Park
State Route 5 & Post Rd., St. Paul, MN 55511—612-727-1961

Southwest of St. Paul, this 3,000-acre park is located at the confluence of the Mississippi and Minnesota Rivers. It has an interpretive center and hiking trails. It is also a good place to experience the riverside wildlife of the area.

Fountain Lake
c/o Albert Lea City Parks, Albert Lea, MN 56007—507-377-4370

Many parks surround this lake in the northwest part of Albert Lea. Facilities include picnicking, boating, and hiking.

Frontenac State Park
Route 2, P.O. Box 134, Lake City, MN 55041—612-345-3401

Located about ten miles southeast of Red Wing on U.S. Route 61, this 2,689-acre park is on the north end of Lake Pepin. It is the gathering place of a wide variety of migratory birds.

Gopher Campfire Game Sanctuary
Hutchinson, MN 55350

Off State Route 7 west of Hutchinson, this sanctuary is home to geese, ducks, deer, and antelope.

Helmer Myre State Park
Route 3, P.O. Box 33, Albert Lea, MN 56007—507-373-5084

Wildflowers are the principle feature of this 1,600-acre park, which is three miles east of Albert Lea. Over 400 species can be found in a heavily wooded island area. Boating, fishing, and hiking are all available. The park is off U.S. 8 along the Minnesota-Wisconsin border.

Heritage Park
225 33rd Ave. South, St. Cloud, MN 56301—612-255-7216

Nature trails and an interpretive center make this a pleasant outing in the city. There is a replica of a working granite quarry at the interpretive center.

Interstate State Park
Taylor Falls, MN 55084—612-465-5711

This 900-acre park was the first interstate park ever established in the United States. Wisconsin and Minnesota were the participating states. Geologic formations, magnificent bluffs, and the enormous potholes that were carved out of the volcanic rock by the meltdown of the great glaciers during the last Ice Age are the main attractions here. One of these potholes is more than 60 feet deep. There are also hiking trails and excursion boats available.

J.C. Hormel Nature Center
21st St., NE, Austin, MN 55912—507-437-7519

An interpretive center, footpaths, woods, ponds, streams, meadows, and prime examples of tall grass prairie are located on the former estate of Jay Hormel. The nature center is open daily, and the interpretive center is open Tuesday–Sunday.

Lac qui Parie State Park
Granite Falls, MN 56241—612-752-4736

Two rivers, dense timberland, boating, fishing, and hiking trails are features of this 529-acre park. It is located 14 miles northwest of Granite Falls on U.S. Route 212.

Lake Pepin

Off State Route 61, Lake Pepin is the widest expanse on the Upper Mississippi River. It contains the largest marina anywhere on the river. The featured wildlife in and around the lake are the many varieties of game fish and the bald eagles that make their homes on the bluffs overlooking the highway. During summer and fall, visitors often see them swooping down into the open water for fish. During the winter, the eagles fish along the shoreline as the ice freezes over the lake.

Lakes Cedar, Harriet, Calhoun, and Lake of the Isles
c/o Minneapolis Department of Parks and Recreation, 310 4th Ave., South, Minneapolis, MN 55415—612-348-2226

Over 12 miles of off-road trails give visitors opportunities to explore the park built around these four lakes. Near the northeast corner of Lake Harriet is the Thomas Sadler Robert Bird Sanctuary, where numerous marshland birds can be observed.

Lakes Hyland, Bush, and Bryant
c/o Hennepin County Park Reserve District, 12615 Rockford Rd., Plymouth, MN 55441—612-559-9000

Southwest of Minneapolis near Bloomington and Eden Prairie, these three lakes are park reserves where naturalists can find easy hiking along developed trails. Visitors may also view a variety of birds and mammals. A nature center and a waterfowl observation tower are both open to the public.

Lilydale–Harriet Island Regional Park
c/o St. Paul Department of Parks and Recreation, City Hall Annex, St. Paul, MN 55101—612-488-7291

Cherokee Bluffs overlook the Mississippi River here. A variety of rocks and fossils can be found in the clay pits near the river.

Lyndale Park
c/o Minneapolis Department of Parks and Recreation, 310 4th Ave., South, Minneapolis, MN 55415—612-348-2226

Off 40th Street and Dupont Avenue on the northeast shore of Lake Harriet sits Lyndale Park. From April to September is the best time to visit this city park with garden displays, exotic and native trees, a rock garden, and a bird sanctuary.

Mille Lacs Kathio State Park
Onamia, MN 56359—612-532-3523

Located eight miles northwest of Onamia on U.S. Route 169, Mille Lacs Kathio is a large, 10,577-acre park that includes the main outlet of one of the largest and loveliest lakes in Minnesota. Much evidence of early Indian habitation dating back at least 4,000 years is featured. Fishing, boating, and hiking are all available.

Minnehaha Park
East 48th St. & Minnehaha Ave., Minneapolis, MN 55417—612-348-2226

Minnehaha Creek flows into the Mississippi River after winding through the western suburbs of Minneapolis. The creek plunges over falls into a forest glen just before reaching the river. Trails follow the creek to the river. This is a popular family park during the summer, and is often crowded and noisy, but some trails lead away from the crowds.

Minneopa State Park
Route 9, Box 143, Mankato, MN 56001—507-625-4388

This park is six miles west of Mankato off U.S. Route 60. Scenic falls and gorge, along with hiking and fishing, are the dominant features of this 1,145-acre park. An added feature is the Minneopa–Williams Outdoor Learning Center, which has a wide variety of native animals and vegetation with information stations. The learning center is open daily from May to October.

Nerstrand Woods State Park
9700 170th St., East, Nerstrand, MN 55053—507-334-8848

This heavily wooded 1,280-acre park has many hiking opportunities. It is 12 miles east Northfield off State Route 246.

Nokomis–Hiawatha Regional Park
Cedar Ave. & Minnehaha Pkwy., Minneapolis, MN 55417—612-348-2226

Two lakes are the features of this city park. Hikers can find plenty of lakeside naturalist's activities.

Pine Grove Park
Little Falls, MN 56345—612-632-2341

West of Little Falls on State Route 27, this park includes a rare tract of virgin white pine. A small zoo with native animals is also located here.

Rice Lake National Wildlife Refuge
Route 2, McGregor, MN 55760—218-768-2402

This 18,140-acre refuge is about 30 miles east of Aitkin near McGregor. Rice Lake, a 4,500-acre migration and nesting area for ducks and geese, is the main feature of the refuge. Walking and automobile trails allow visitors to view the large flocks of ducks and geese that use the area as a stopover along the Mississippi Flyway.

River Bend Nature Center
Rustad Rd., Faribault, MN 55021—507-332-7151

Ten miles of trails meander through the 640 acres of mixed habitat that include woodlands, prairie, ponds, and rivers. The center is about one mile southeast of town.

Savanna Portage State Park
HCR 3, Box 591, McGregor, MN 55760—218-426-3271

The park is about 35 miles northeast of Aitkin off U.S. Route 210 on State Route 65. It is a wilderness area built around the historic portage that linked the Mississippi River and Lake Superior. The 15,818 acres feature wilderness hiking, boating, and fishing.

Sherburne National Wildlife Refuge
Route 2, Zimmerman, MN 55398—612-389-3323

A wide variety of wildlife can be observed in this refuge 13 miles north of Elk River off State Route 169. A wildlife observation center, interpretive hiking trails, and self-guided auto tours all offer visitors opportunities to view mammals and birds in their natural habitats.

Sibley Park
Park Ln. & Givens St., Mankato, MN 56001—507-625-3165

This is a 100-acre city park with a river walk, gardens, and scenic views. The Blue Earth and Minnesota Rivers join here, and the riversides are heavily wooded.

Sibley State Park
Willmar, MN 56201—612-354-2055

About 15 miles north of Willmar on U.S. Route 71, this was the favorite hunting ground of the state's first governor, for whom it is named. The 2,300-acre park has a variety of activities, including a nature center and hiking trails.

St. Anthony Falls
Main St. SE & Central Ave., Minneapolis, MN 55414

This is not truly a nature outing. A public vantage point here gives visitors a view of the upper locks, dams, and falls that mark the head of the navigable portion of the Mississippi River.

St. Croix and Lower St. Croix National Scenic Riverway
P.O. Box 708, Taylor Falls, MN 55084—612-483-3284

Two segments of this river, totaling more than 250 miles, have been designated as National Scenic Riverways, and are administered by the National Park Service. The river flows southward from northern Wisconsin, and forms the Minnesota–Wisconsin border until it joins the Mississippi River near Point Douglas. Information about the river and National Park Service facilities can be obtained by calling or writing the St. Croix National Scenic Riverway.

St. Croix State Park
Route 3, P.O. Box 450, Hinckley, MN 55037—612-384-6591

This is a very large park, 34,037 acres, with a wide variety of activities, including hiking trails, lakes, streams, and an interpretive center. It is 15 miles southeast of Hinckley off State Route 48.

Theodore Wirth Park
and
Eloise Butler Wildflower Garden and Bird Sanctuary
Glenwood Ave. & Xerxes Ave., North, Minneapolis, MN 55422—612-348-2226

Wooded hills are the feature of this park. Many miles of trails and roads are favorites for hiking and cross-country skiing enthusiasts. The 20-acre wildflower garden and bird sanctuary contain a natural bog and swamp, plus native prairie and woodland flower and bird habitats. These facilities are open daily from April to October. A free, self-guided tour book is available for visitors.

University of Minnesota Landscape Arboretum
3675 Arboretum Dr., Chanhassen, MN 53317—612-443-2460

Six miles of trails crisscross the 675 acres of hills, lakes, and marshland with hundreds of tree and shrub varieties. This is a good bird-watching area. The grounds are open daily from 8:00 A.M. to sunset. The arboretum is off State Route 5 west of Chanhassen.

Upper Sioux Agency State Park
Granite Falls, MN 56241—612-564-4777

This 1,280-acre park has an interpretive center, hiking, and boating. It is eight miles southeast of Granite Falls on State Route 67.

Wild River State Park
Taylor Falls, MN 55084—612-583-2125

This 6,706-acre park has over 50 miles of hiking and horseback riding trails, rental canoes, an interpretive center, and a trail center. It is located in St. Croix River Valley about ten miles northwest of Taylor Falls off State Route 95.

William O'Brien State Park
Stillwater, MN 55082—612-433-2421

This medium-sized, 1,273-acre park is 16 miles north of Stillwater on State Route 95. It has hiking and interpretive programs, along with the standard facilities for camping, fishing, and boating.

Organizations That Lead Outings

Audubon Society
1313 SE 5th St., Minneapolis, MN 55414—612-379-3868

Friends of the Boundary Waters Wilderness
1313 SE 5th St., Minneapolis, MN 55414—612-379-3835

Izaak Walton League of America
6601 Auto Club Rd., Bloomington, MN 55438—612-941-6654

Sierra Club
1313 SE 5th St., Minneapolis, MN 55414—612-379-3853

A Special Outing

The Mississippi River bisects Minneapolis and St. Paul from north to south. It is more famous than the Minnesota River, a tributary of the Mississippi. However, it is the Minnesota that forms the southern boundary of the Twin Cities and offers visitors more naturalist opportunities than the Mississippi.

The Minnesota flows through a 250-foot deep and five-mile wide trench that is the former riverbed of an ancient glacial river. Tree-lined bluffs extend as far as you can see along the edge of this ancient gorge. The present river meanders through floodplains formed centuries ago.

The gorge forms a natural moat south of Minneapolis–St. Paul that is breeched by a series of bridges. It is easy to imagine that a landscape painting from the 19th century has come to life as you scan the valley where the Minnesota flows. A quick glance around assures you that time has not turned backward, for smokestacks and grain elevators stand silhouetted against the horizon.

The marshes that align the river have never been drained and developed as the metropolitan area grew around them. This is so for a variety of reasons, but primarily because of regular flooding that occurs on the floor of the valley.

As early as the 1930s Thomas Wirth, the first superintendent of the Minneapolis Department of Parks and Recreation, wanted to save the floodplains in their natural state. But it was not until the late 1960s, when two large floods did extensive damage along both the Mississippi and Minnesota rivers, that action was taken to protect the gorge.

After extensive lobbying efforts by several citizen groups, an urban national wildlife refuge was formed to protect 7,000 acres of the Minnesota River Valley. Along with another 4,000 acres of state-owned lands, including Fort Snelling State Park that adjoins the Minnesota Valley National Wildlife Refuge, a 36-mile stretch of the Minnesota River and adjacent marshland, have been preserved in their natural state. These parks and refuges fit well with the other parks in the Minneapolis–St. Paul area. They offer visitors opportunities to explore a protected riparian environment within easy reach of downtown.

The woods around Bass Lake and the meadows at Louisville Swamp are not exactly escapes from civilization, but they are home to an astounding variety of wildlife. In Long Meadow Lake muskrats and blue herons live side by side, unhindered by their proximity to an urban area.

Also featured in the Minnesota River Valley are rare fens. These are wet, peat-filled places with low-lying hammocks. They are formed where groundwater bubbles to the surface and keeps the soil moist. These areas tend to have a convex-lens shape and are elegant at close range, if nondescript at a distance.

The fens in the Minnesota Valley are a special kind. There the underground water is rich in calcium and magnesium bicarbonates. These minerals leave deposits that determine which plants survive in the alkaline environment of the hammocks.

Calcareous fens are thinly scattered over the glacial zone of North America, and five survive in the lower end of the Minnesota Valley. Others were destroyed over the years before botanists became aware of the special characteristics of the flora found on them. The calcium-tolerant rushes, grasses, asters, gentians, and lady slippers are all remnants of the Ice Ages.

None of the remaining fens are completely within the boundaries of the Minnesota Valley National Wildlife Refuge. One is protected by Fort Snelling

State Park, which is contigious with, and for all practical purposes, part of the refuge. Another is nearby and is owned by the Minnesota Department of Natural Resources. A third, Nicols Meadowland, is privately owned.

The largest is Savage Fen, which is over 500 acres. A variety of owners control it. The Nature Conservancy owns about 30 acres. Twenty-six acres have been donated to the refuge, and the rest is privately owned.

Rare Ice Age plants live on the calcareous fens of the Minnesota Valley, Beavers, muskrats, and other marshland mammals hunt and build their homes in the hammocks and low-lying land of the valley floor. Numerous birds and waterfowl feed, winter, and nest along the waterways and lakes of the region. All these combined offer visitors to the Minnesota Valley Wildlife Refuge a great chance to explore wilderness nearby.

Nature Information

Hennepin County Parks Department
12615 Rockford Rd., Plymouth, MN 55441—612-559-9000

Minneapolis Chamber of Commerce Tourism Department
15 S. 5th St., Minneapolis, MN 55402—612-370-9132

Minneapolis Department of Parks and Recreation
310 4th Ave., South, Minneapolis, MN 55415—612-348-2226

Minnesota Department of Natural Resources
500 Lafayette Rd., P.O. Box 40, St. Paul, MN 55146—612-296-6157

Minnesota Tourism Division
250 Skyway Level, 375 Jackson, St. Paul, MN 55101—612-296-5029

St. Paul Chamber of Commerce
600 N. Central Tower, St. Paul, MN 55101—612-222-5561

St. Paul Department of Parks and Recreation
City Hall Annex, St. Paul, MN 55101—612-488-7291

Further Reading

Blacklock, Les, and Craig Blacklock. *Minnesota Wild.* Stillwater, MN: Voyageur Press, 1983.

Buchanan, James W. *Minnesota Walk Books.* Minneapolis: Nodin Press, 1978.

Coffin, Barbara, and Lee Pfannnmuller. eds. *Minnesota's Endangered Flora and Fauna.* Minneapolis: University of Minnesota Press, 1983.

Daniel, Glenda, and Jerry Sullivan. *A Sierra Club Naturalist's Guide to the North Woods of Michigan, Wisconsin, Minnesota, and Southern Ontario.* San Francisco: Sierra Club Books, 1981.

Hazard, Evan B. *The Mammals of Minnesota.* Minneapolis: University of Minnesota Press, 1982.

Janssen, Robert B. *Birds in Minnesota.* Minneapolis: University of Minnesota Press, 1987.

Shepard, John G. *Minnesota.* Chester, CT: The Globe Pequot Press, 1989.

St. Louis

St. Louis sits on the banks of the Mississippi River near the middle of the eastern border of Missouri at the head of the Ozark Uplift. The Ozarks run southwest from the city into Arkansas and Oklahoma. The highest point in the chain is Taum Sauk Mountain, which is about 90 miles south of St. Louis.

Naturalist activities in and around St. Louis include activities on the Mississippi River, along other nearby rivers in Missouri, and in the rolling woodlands of Missouri and southern Illinois.

St. Louis has another popular activity that is almost nonexistent in other areas; that is cave and cavern exploring. The eroded Ozark highlands have many caves that have been carved by water over the centuries. Some of these are noted in the listings as easy outings for visitors who have never explored caves.

ABOVE: Visitors to Illinois' Pere Marquette State Park can walk on hillsides covered with mixed hardwood forests that overlook the Mississippi River. Photo courtesy of Illinois State Park Service.

Of the other cities in this guide, only Cleveland offers this opportunity, and caves are much more limited there.

Without a doubt the Mississippi River remains the dominant natural feature of St. Louis. Yet little of the river and the surrounding land has been preserved in a natural state for visitors or residents to enjoy. More of the bottomlands of smaller rivers and creeks have been made into parks and preserves as you get farther away from the Mississippi. Several of the entries for southern Illinois include prime examples of bottomland woods.

Most of the region around St. Louis is covered by mixed hardwood or mixed bottomland forests. To the south and west of the city lie the Ozarks, with mixed hardwood and pine forests cirsscrossed by numerous creeks and rivers. Many of these have been dammed to form large human-made lakes.

Climate and Weather Information

Based on a number of factors, including climate, in terms of livability, St. Louis has been rated as high as seventh in the United States. Visitors to the city during one of its extremely hot and humid summers, or cold and windy winters may find this difficult to believe. Summer temperatures can reach between 90° and 100°F, and the humidity may peak between 80 and 100 percent. During the winter, temperatures may fall down to 0°F. These periods of extreme temperatures seldom last long, however, and are replaced by mild weather that is conducive to outdoor activities.

Summers are warm, with some periods of hot weather, which are aggravated by the humidity from the Mississippi and other nearby rivers. Winters can be brisk, while spring and fall are generally mild. St. Louis residents enjoy the long falls, which stretch from September into late November and warm springs, when flowering bulbs, shrubs, and wildflowers bring a mass of color to the countryside. Rain is likely to come at any time from spring through summer, and visitors may encounter snow in January and February.

Getting Around St. Louis

The streets in downtown generally follow a basic grid, but are complicated because of many one-way streets. On-street parking is hard to find, and off-street parking rates are average to low. Rental cars agencies are numerous and taxis are easy to find. Public transportation serves all of the city of St. Louis as well as St. Louis and Jefferson Counties. You can get to many of the activities listed by public transportation, but will need to rent a car to get to outlying sites.

Indoor Activities

Bonne Terre Mine Tours
Park & Allen Sts., Bonne Terre, MO 63628—314-358-2148

These caverns were mined for lead and silver from 1870 to 1962. They are larger than the town of Bonne Terre. Tours give visitors a view of old mining tools, a flower garden, and a billion-gallon underground lake. The mine is open daily from 9:00 A.M. to 6:00 P.M.

Caverns

Many caverns can be visited during inclement weather. See the Outdoor Activities section for details.

Ed Clark Museum of Missouri Geology
Fairground Road in Buehler Park, Rolla, MO 65401—314-364-1752

These displays of Missouri's minerals, fossils, and geological history are located at the Department of Natural Resources, Division of Geology and Land Survey. The museum is open Monday–Friday, 8:00 A.M. to 5:00 P.M.

Jewel Box Floral Conservatory
Wells & McKinley Drives, St. Louis, MO 63119—314-535-5050

This greenhouse has floral displays and special holiday shows. It is open daily from 9:00 A.M. to 5:00 P.M.

Minerals Museum
Norwood Hall, University of Missouri, 12th & Rolla Sts., Rolla, MO 65401—314-341-4801

This collection began with the Missouri mining exhibit for the St. Louis World's Fair in 1904. It is now one of the best collections in the state. Over 4,000 specimens from 92 countries and 47 states are on exhibit. The museum is open Monday–Friday from 8:00 A.M. to 5:00 P.M.

St. Louis Science Center
5100 Clayton Ave., St. Louis, MO 63110—614-289-4400

This science center is built around three major themes—earth science, life science and anthropology, and science and technology. It has more than 50 exhibits, including live and hands-on displays. A planetarium with computer generated special effects is also in the building. The center is open daily from 10:00 A.M. to 5:00 P.M.

University of Missouri
Visitor and Guest Relations, 103 Heinkel Bldg., Columbia, MO 65211—314-882-6333

A variety of naturalist-oriented museums and collections are located on campus. Among them are the following.

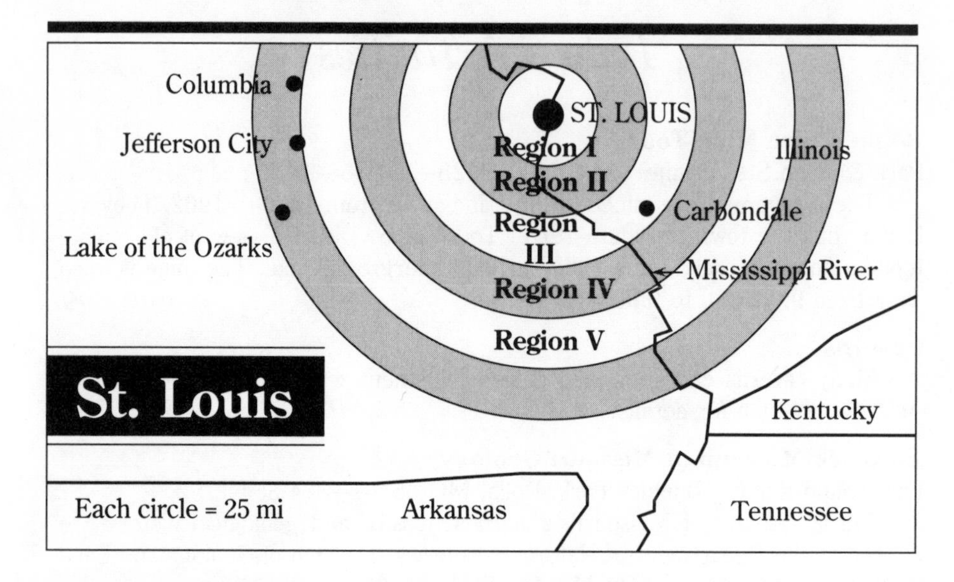

Indoor Activities

Region I
Jewel Box Floral Conservatory
St. Louis Science Center

Region II
Bonne Terre Mine Tours

Region IV
Ed Clark Museum of Missouri Geology
Minerals Museum
University of Missouri

Outdoor Activities

Region I
Edgar M. Queeny Park
Laumeier Sculpture Garden
Mastodon State Park
Missouri Botanical Gardens
Shaw Arboretum
Wolf Sanctuary

Region II
Dr. Edmund A. Babler Memorial State Park
Lone Elk
Pere Marquette State Park
Washington State Park

Region III
Elephant Rocks State Park
Graham Cave State Park
Johnson's Shut-ins State Park
Meramec Caverns
Meramec State Park
Onondaga Cave State Park
Sam A. Baker State Park
Taum Sauk Nature Museum
Taum Sauk Pumped Storage Plant

Region IV
Little Dixie Lake Wildlife Area
Mitchell Museum
Montauk State Park
Stephen A. Forbes State Park

Region V
Beall Woods State Park
Big Oak Tree State Park
Chautauqua National Wildlife Refuge
Clearwater Lake
Ha Ha Tonka State Park
Newton Lake State Fish and Wildlife Area
Ozark National Scenic Riverways
Shawnee National Forest
Shelbyville Fish and Wildlife Area
Shelter Gardens
Trail of Tears State Park
Walnut Point State Fish and Wildlife Area

The Geology Building has more that 100,000 fossil and rock specimens, many of which are on display in the corridors of the ground and first floor. Lefevre Hall has a large collection of mounted waterfowl. Both are open Monday–Friday from 8:00 A.M. to 10:00 P.M.

In Laws Observatory visitors can look through the 16-inch telescope on clear Friday nights between 8:00 P.M. and 10:00 P.M. when classes are in session. In Stewart Hall Room 217, visitors can watch the mid-morning feeding of reptiles on Fridays.

The Botany Greenhouse and Herbarium has special rooms of tropical, jungle, and desert flora. It is open Monday–Friday from 8:00 A.M. to 5:00 P.M. Phone 314-882-6888. Just east of the Agriculture Building lie small woodland and floral gardens duplicating much of the state's landscapes.

Outdoor Activities

Beall Woods State Park

R.R. 2, Mt. Carmel, IL 62863—618-298-2442

The park is six miles south of Mt. Carmel, Illinois, off State Route 1. These 635-acres were purchased by the State of Illinois in 1965 to preserve one of the largest single tracts of deciduous forest in the United States. It has remained largely untouched by humans. Explorers and naturalists such as John Audubon, George Clark, and Robert Ridgeway all visited this region.

Beall Woods is a natural ecological forest system that is also known as the "University of Trees" because of its variety. A number of hiking and interpretive trails lead through the woods and give visitors opportunities to observe the native flora and fauna of the region.

Big Oak Tree State Park

East Prairie, MO 63845—314-649-3149

Some of the largest trees in the country can be found in this park. A one- and one-quarter-mile boardwalk winds through virgin bottomland hardwoods that are home to more than 140 species of birds, plus other abundant wildlife. The park is about 12 miles east of East Prairie, Missouri, on State Route 102.

Chautauqua National Wildlife Refuge

Route 2, Havana, IL 62644—309-535-2290

This refuge covers 4,488 acres of Illinois River bottomland. As many as 150,000 waterfowl have been observed at one time during the fall migration. Bald eagles can be seen during the fall and winter. A nature trail and observation tower give visitors a good view of the refuge.

Clearwater Lake

P.O. Box 68, Piedmont, MO 63957—314-223-7777

An Army Corps of Engineers project, Clearwater Lake was formed by a dam on the Black River. In addition to boating and fishing, there are nature and exercise trails around the lake. It is about 20 miles northeast of Van Buren, Missouri, off U.S. Route 60.

Creve Coeur Park
Marine & Dorsett Rds., St. Louis, MO 63118—314-889-2894

Hiking, fishing, canoeing, and sailing are all offered at this 1,141-acre park. It surrounds 320-acre Creve Coeur Lake.

Dr. Edmund A. Babler Memorial State Park
800 Guy Park Dr., Chesterfield, MO 63005—314-458-3813

This park is about 30 miles west of St. Louis off State Route 100. The 2,439-acre facility features an interpretive center, naturalist, and hiking trails.

Edgar M. Queeny Park
550 Weidman Rd., St. Louis, MO 63011—314-391-0900

This is a medium sized, 569-acre, regional park with hiking trails. It is located 20 miles north of St. Louis on State Route 100.

Elephant Rocks State Park
Pilot Knob, MO 63663—314-697-5395

This small, 129-acre park has interesting billion-year-old granite formations. Informational trails lead through the formations. The park is nine miles north of Pilot Knob, Missouri on State Route 21.

Graham Cave State Park
Hermann, MO 65041—314-564-3476

This 357-acre park has a sandstone cave that Indians occupied as long as 10,000 years ago. There are also nature trails and camping. The park is off Interstate 70, about 20 miles north of Herman, Missouri.

Ha Ha Tonka State Park
Camdenton, MO 65020—314-346-2986

About five miles southwest of Camdenton, Missouri, on U.S. Route 54, this 2,481-acre park is on the Niangua Arm of the Lake of the Ozarks. Its most significant natural features can be found near the 750-acre area where the lake joins with the Ha Ha Tonka Spring. All the unique features are the remnants of an immense ancient cavern system. Features include a natural pit, a 150-foot deep sink basin, and a 100-foot high and 70-foot wide natural bridge. The entire area is a classic example of a geological topography known as *karst*, or an irregular limestone region with sinks, underground streams, and caverns.

Horseshoe Lake Wildlife Refuge
Illinois Route 111, Granite City, IL 62040—618-931-0270

This 2,000-acre refuge around a lake named for its shape is a stopping place for over 100,00 Canada geese during the fall migration. The lake is also a popular fishing site.

Johnson's Shut-in's State Park
Pilot Knob, MO 63663—314-546-2450

Follow State Route 21 four miles north of Pilot North, Missouri. Then go 14 miles southwest on County Road H.

The 20-mile Taum Sauk backpacking trail, which leads to Pilot Knob and over Missouri's highest mountain, Taum Sauk, begins in this park. The 2,430 acres in this rugged wilderness include a spectacular canyon defile along a river and several hiking trails.

Laumeier Sculpture Garden
Geyer & Rott Rds., St. Louis, MO 63127—314-821-1209

This outdoor sculpture park on the grounds of the Laumeier mansion features nature trails and a picnic area.

Little Dixie Lake Wildlife Area
Fulton, MO 65251—314-642-3816

Located ten miles northwest of Fulton, Missouri, this area features lake activities, fishing, and bird watching. The concession area has food, bait, tackle, and boat rentals from April through September.

Lone Elk
State Route 141 & North Outer Rd., St. Louis, MO 63141—314-889-3192

This is a 405-acre preserve for bison, elks, deer, and Barbados sheep. It is about 25 miles southwest of St. Louis on Interstate 44. The preserve is open daily from 8:00 A.M. to sunset.

Mastodon State Park
1551 Seckman Rd., Imperial, MO 63052—314-464-2976

This park is 20 miles south of St. Louis off U.S. Route 67 at the Imperial Exit. Ongoing excavation of American mastodon remains and Indian artifacts are the main attractions here. There are also hiking trails in this day-use park.

Meramec Caverns
Sullivan, MO 63080—314-468-3166

Use Exit 230 off Interstate 44 south of Sullivan, Missouri. This cavern has five levels, and was once used by Jesse James as a hideout. Guided tours are given daily from March to December.

Meramec State Park
Sullivan, MO 63080—314-468-6072

This park is four miles east of Sullivan, Missouri on State Route 185. It is one of Missouri's largest parks, with 6,489 acres. It has over 20 caves, many springs, a nature museum, naturalist program, and hiking trails.

Missouri Botanical Gardens
4344 Shaw Blvd., St. Louis, MO 63110—314-577-5100

This is a 79-acre garden. It includes the world's first geodesic dome greenhouse, the largest traditional Japanese garden in America, and a four-acre lake with two waterfalls and four islands. One of the oldest and most famous botanical gardens in the world, it is often compared to London's Kew Gardens.

Mitchell Museum
Richview Dr., Mt. Vernon, IL 62864—618-242-1236

While this is a fine arts museum, there is a bird sanctuary around Mitchell Lake, which is on the museum grounds. This sanctuary includes the Juniper Ridge and Cedarhurst Braille Trails. There is also wheelchair access.

Montauk State Park
P.O. Box 38, Montauk, MO 63039—314-548-2201

Trout fishing and hiking trails are the main attractions of this 1,193-acre park. It is about 40 miles south of Rolla, Missouri, off U.S. Route 63.

Newton Lake State Fish and Wildlife Area
R.R. 4, P.O. Box 178B, Newton, IL 62448—618-783-3478

White-tailed deer, blue herons, ducks, and geese are only some of the wildlife that hikers see along the Newton Lake trails. The area also has one of the three largest flocks of prairie chickens in Illinois. These can also be found in the adjacent 240-acre refuge that is run by the Central Illinois Public Service Company.

Over 20 miles of marked hiking trails are in this 2,315-acre refuge. Fishing is allowed on the 1,775-acre lake. The area is off State Route 33 between Dietrich and Newton, Illinois.

Onondaga Cave State Park
Sullivan, MO 63080—314-468-6072

An historical cave site is the main feature of this 1,300-acre state park. Canoe trips on the nearby river are also an attraction. Onondaga Cave State Park is about 15 miles southwest of Sullivan, Missouri, off Interstate 44.

Ozark National Scenic Riverways
P.O. Box 490, Van Buren, MO 63965—314-323-4236

This area is off U.S. Route 60 near Van Buren, Missouri. The National Park Service administers more than 80,000 acres along the Current and Jacks Fork Rivers. These clear-flowing rivers are fed by a series of springs that make the region famous. The largest of these is Big Spring, which has a flow of over 276,000,000 gallons a day. Caves, riparian wildlife habitat, fishing, and river floating are all featured in the park.

Pere Marquette State Park
P.O. Box 158, Grafton, IL 62037—618-786-3323

Located five miles west of Grafton, Illinois, on State Route 100, this 7,500-acre state park is Illinois' largest. It is at the confluence of the Illinois and Mississippi Rivers. An interpretive center, hiking trails, fishing, boating, and handicapped facilities are all featured.

Sam A. Baker State Park
Patterson, MO 63956—314-856-4411

This is a large, 5,168-acre, state park that includes the St. Francois and Big Creek Rivers. These provide the main attractions. Trail and wilderness hiking are also available. The park is about 40 miles south of Pilot Knob, Missouri, off State Routes 21 and 143.

Shaw Arboretum
P.O. Box 38, Gray Summit, MO 63039—314-742-3512

This arboretum contains four square miles of natural Ozark landscape. It is crisscrossed by ten miles of trails.

Shawnee National Forest
Interstate 45 South, Harrisburg, IL 62946—618-253-7114

This 262,500-acre national forest extends from the Mississippi River east to the Ohio River. It offers a sharp contrast to the flat farmland to the north. The region is filled with hills and valleys, wildlife, unusual rock formations, and scenic vistas. The forest is easily accessible and has many hiking trails.

The Rim Rock Forest Trail winds past some of Illinois' best geological sites. The Garden of Gods Trail explores a 200-million-year-old rock formation and has an interpretive center. The Bell Smith Springs Trail features caves and a natural bridge. Maps and more information are available from the Forest Supervisor.

Shelbyville Fish and Wildlife Area
R.R. 1, P.O. Box 42-A, Bethany, IL 61914—217-665-3112

This area is along the Kaskaskia and West Okaw Rivers between Shelbyville and Sullivan, Illinois. There are three separate units in this area: the 2,700-acre West Ocraw Unit; the 3,700-acre Kaskaskia Unit; and the 960-acre Eagle Creek Wildlife Area. An 11,000-acre lake has dozens of islands. Protected bays and coves provide ideal settings for watching waterfowl.

The upland shores of Lake Shelbyville are covered with mixed hardwood forests, and the lowland forests have cottonwood, soft maple, willow, and sycamore. Both forests feature colorful wildflowers in the spring and summer. Prairie grasses have reclaimed many of the abandoned fields. Hiking trails have been built in all three units, and fishing is excellent in the creeks and ponds.

Shelter Gardens
1817 W. Broadway, Columbia, MO 65203—314-445-8441

This miniature midcontinent environment is located on the grounds of the Shelter Insurance Company. A pool, stream, domestic and wildflowers, and more than 300 tree and shrub species are featured. There is also a garden for the blind.

Stephen A. Forbes State Park
R.R. 1, Kinmundy, IL 62854—618-547-3381

This is a 3,100-acre park with fishing, hiking, horseback riding, and camping. It is about 15 miles northeast of Salem, Illinois, off U.S. Route 50.

Taum Sauk Nature Museum
and
Taum Sauk Pumped Storage Plant
Hogan, MO 63650—314-637-2281

The plant and museum are ten miles northwest of Hogan, Missouri, off State Route 21. The power plant produces 350,000 kilowatts of electricity by a unique process. A 55-acre reservoir sits above the power plant, and water drops about 7,000 feet through a tunnel to the plant. It then flows into a 400-acre holding lake. At night, during periods of low electrical use, water is pumped back to the upper reservoir to be reused the following day.

An observation platform gives visitors an opportunity to view how the plant operates. The adjoining nature museum depicts the natural resources of Missouri.

Trail of Tears State Park
Cape Girardeau, MO 63701—314-334-1711

This 3,306-acre park sits on limestone bluffs that overlook the Mississippi River. It commemorates the forced movement of the Cherokee Indians over the Trail of Tears to Oklahoma. Fishing, boating, and hiking are all available. This park is ten miles north of Cape Girardeau, Missouri, on State Route 177.

Walnut Point State Fish and Wildlife Area
Route 2, Oakland, IL 61943—217-346-3336

This area is 20 miles northeast of Charleston, Illinois, between State Routes 36 and 133. Spring brings wildflowers and redbuds to this 464-acre area. Hikers often catch a glimpse of foxes, raccoons, turtles, and snakes that are native to this area. Many varieties of birds can be seen at various times during the year. Lake and river fishing are both available.

Washington State Park
Bonne Terre, MO 63628—314-586-2995

Hundreds of Indian petroglyphs can be viewed in this 1,415-acre park. Canoeing on the Big River, hiking trails, and a nature museum are other attractions. The park is 15 miles northwest of Bonne Terre, Missouri, on State Routes 47 and 21.

Wolf Sanctuary
P.O. Box 760, Eureka, MO 63025—314-938-5900

This project of Carole and Marlin Perkins, of "Wild Kingdom" fame, is located at the Tyson Valley Research Center of Washington University. It is devoted to the preservation of the many varieties of wolves that roam freely in packs at the facility. Visitors can join morning tours from July to January. No tours are offered during the spring breeding and whelping season.

Organizations That Lead Outings

The Nature Conservancy
2800 S. Brentwood Blvd., St. Louis, MO 63117—314-968-1105

The Sierra Club
1005 S. Big Bend Blvd., St. Louis, MO 63117—314-645-1019

Wolf Sanctuary
P.O. Box 760, St. Louis, MO 63025—314-938-5900

A Special Outing

Even though it has an abundance of water, Missouri has few natural lakes. This was long a problem for residents, as steady, strong rainfall often brought flooding to the lowlands next to the many rivers and creeks that crisscross the state. While this was a natural occurrence, the people whose homes were flooded on a regular basis put pressure on state and federal governments to control the flooding.

As was normal in the earlier part of this century, various agencies responded to this request. Numerous flood control dams were built along many tributaries of the Mississippi River that flow through Missouri. Reservoirs now cover thousands of bottomland acres.

This bottomland, which was home to a wide variety of flora and fauna, has been lost. Visitors to Missouri can utilize the large lakes that have been impounded by these dams, however, to explore some of the Ozark hill country that was difficult to reach before the dams were built.

Today reservoirs, such as the huge Lake of the Ozarks, have boat facilities. Visitors can rent boats and use them to cross over isolated arms of the reservoirs to explore almost undisturbed upland mixed forests where wildflowers bloom profusely in the spring, and birds nest in peace.

For more information about the human-made lakes in Missouri, and the activities in and around them, contact either the Missouri Department of Natural Resources or the Missouri Division of Tourism. The addresses are shown in the next section.

Nature Information

Missouri Department of Natural Resources
Division of Parks, P.O. Box 176, Jefferson City, MO 65102—314-751-2479

Missouri Department of Conservation
P.O. Box 180, Jefferson City, MO 65102—314-751-4115

Missouri Division of Tourism
P.O. Box 1055, Jefferson City, MO 65102—314-751-4133

National Forests in Missouri
U.S. Forest Service, 401 Fairgrounds Rd., Rolla, MO 65401—314-364-4621

St. Louis Convention and Visitors Commission
10 S. Broadway, St. Louis, MO 63102—314-421-2100 for recorded message, 314-421-1023 for local calls, 800-325-7962 for out-of-state calls

Further Reading

Beveridge, Thomas R. *Geologic Wonders and Curiosities of Missouri.* Rolla, MO: Missouri Division of Geology and Land Survey. Educational Series No. 4, 1978.

Boyer, Chris. *Missouri Parks Guide.* Wauwatosa, WI: Affordable Adventures, 1988.

Denison, Edgar. *Missouri Wildflowers.* 2nd Ed. Jefferson City, MO: Missouri Department of Conservation, 1973.

Knittel, Robert E. *Walking in Tower Grove Park: A Victorian Strolling Park.* 2nd Ed. St. Louis: Grass Hopper Press, 1984.

Lafser, Fred A., Jr. *A Complete Guide to Hiking and Backpacking in Missouri.* Annapolis, MO: Fred A. Lafser, Jr., 1974.

Loughlin, Caroline, and Catherine Anderson. *Forest Park.* Columbia, MO: University of Missouri Press, 1986.

Pflieger, William L. *The Fishes of Missouri.* Jefferson City, MO: Missouri Department of Conservation, 1975.

Rafferty, Milton D. *The Ozarks Outdoors.* Norman, OK: University of Oklahoma Press, 1985.

———. *Missouri: A Geography.* Boulder, CO: Westview Press, 1983.

Wylie, J.E., and Ramon Gass. *Missouri Trees.* Jefferson City, MO: Missouri Department of Conservation, 1973.

Zimmerman, John L., and Sebastian T. Patti. *A Guide to Bird Finding in Kansas and Western Missouri.* Lawrence, KS: University of Kansas Press, 1988.

SOUTHWEST

Houston
Dallas
Denver
Phoenix

Of all the regions discussed in this guide, the natural history of the four cities of the Southwest are most dissimilar. Houston has more in common with New Orleans, with its bayous, water, and coastal-like environment. Dallas sits on the prairie astride the dividing line between the eastern and western United States. Denver is at the junction of the Great Plains and the Rocky Mountains, and Phoenix is archetypical of the southwest desert.

With such diversity, it is difficult to make general statements about their natural history. One statement that can be made, though, is that all these cities experienced explosive growth after World War II, much of which was uncontrolled. In most instances, this rapid expansion damaged the natural environment in and around the cities, as it destroyed fragile and irreplaceable habitats.

Natural History

Houston sits amidst the swampland of the Gulf Coastal Plain. It offers visitors opportunities to observe some of the highest concentrations and widest variety of birds in the United States. This is particularly true during the spring and fall when migration is in full swing.

In addition, Houston offers its visitors access to the Big Thicket country northeast of the city. This is the southern extension of 26 million acres of pine, oak, and cypress woodlands that border the Sabine River in east Texas. These nearly impenetrable forests are one of the last remnants of a vanishing habitat.

Dallas lies in what is called the Central Lowlands section of the state where there are rolling hills and fertile plains. Denver sits at the junction of the Great Plains and the Rocky Mountains. It offers visitors opportunities to explore a wide variety of naturalist outings in both mountains and plains.

Phoenix offers desert, as visitors move from the flatlands of the valleys around Phoenix to the surrounding mountains. Even as you venture to Tucson, you do not leave the desert until the upper reaches of the Santa Catalina Mountains to the north of the city.

Houston

Houston is an excellent example of how humans have attempted to control nature and change the natural environment to meet the needs of an industrial society. Although it sits more than 50 miles inland, Houston is one of the busiest seaports in America.

The city was built as a maritime trading post on the winding, marshy Buffalo Bayou during the 1830s. It grew into an important trading center as cotton became king by the end of the 19th century. Houston became even more important after the discovery of oil in the region at the beginning of the 20th century. As Houston reached boomtown status, Buffalo Bayou was converted into the Houston Ship Canal. This conversion precipitated even more rapid industrial growth.

In the 20 years between 1940 and 1960 Houston grew from the 27th largest city in the United States to the sixth largest. It is presently the fourth largest.

ABOVE: Visitors can explore a variety of forest habitats in the Big Thicket National Preserve. Photo courtesy of Big Thicket National Preserve.

During this growth Houston engulfed large chunks of marshlands that were once home to a wide variety of wildlife.

Fortunately, even such rapid growth has not completely eradicated the vast marshlands south and east of the city. Visitors can still explore miles of habitat that extend all the way to the Gulf of Mexico. Only California rivals Texas for the number of bird species that live in it, or pass through annually, and Houston sits in the mainstream of the Great Central Flyway.

Climate and Weather Information

Houston has a warm-to-hot climate where the temperatures seldom fall below freezing. The fine gulf mists that frequently fill the air, along with the high temperatures, combine to ensure that Houston experiences humid summers.

Spring and fall are long, pleasant seasons, and winter is seldom uncomfortable. Summer outings are often uncomfortable, not only because of the heat and humidity, but also because of the large numbers of biting insects. Rain is frequent in the summer, and the city is occasionally hit by hurricanes.

Getting Around Houston

Residents call Houston's freeway system the "Spaghetti Bowl" because of its complexity. Major routes spiral out from the central area, not following any east-west or north-south format. Their names often change with a change of direction. All of this confuses many visitors. On-street parking is almost nonexistent downtown, but commercial lots and garages charge reasonably priced rates.

Public transportation is adequate in the city and to the suburbs. Rental car agencies are found in most major hotels, and taxis are plentiful, but the fares are somewhat expensive. Most of the outings listed are away from the city center, and you will need to rent a car to see them.

Indoor Activities

Museum of Natural Science, Museum of Medical Science, and Burke Baker Planetarium

1 Hermann Circle Dr., Houston, TX 77030—713-526-4273

These three museums are all located in the same building in Hermann Park. The Museum of Natural History is a place for doing things. Touch displays, movies, a library, and classes are all available to visitors. A large exhibit on petroleum science is located on the first floor. There are ever-changing dioramas on the second floor. When new dioramas are constructed visitors can see the artists at work. The planetarium offers five different shows yearly. Call for current

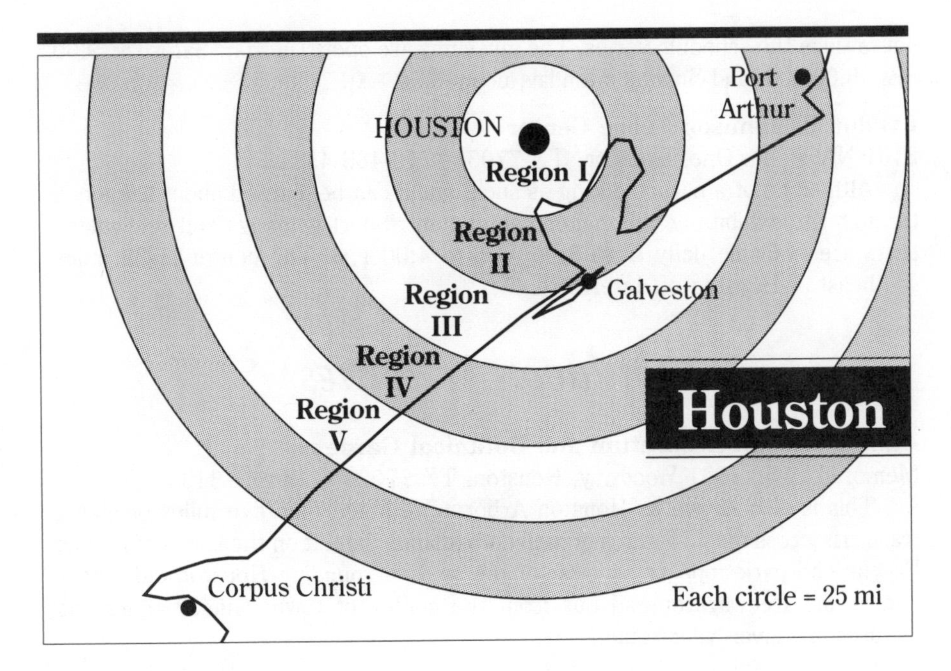

Indoor Activities

Region I

Museum of Natural Science, Museum of Medical Science, and Burke Baker Planetarium

Region II

Lyndon B. Johnson Space Center

Outside Activities

Region I

Aline McAshan Arboretum and Botanical Gardens
Armand Bayou Nature Center
Bay Area Park
Edith L. Moore Nature Sanctuary
Houston Garden Center
Houston Hiking Trails
Memorial Park
Mercer Park Arboretum

Region II

Anahuac National Wildlife Refuge
Big Creek Recreation Area
Brazos Band State Park
Double Lake Recreation Area
Galveston Island State Park
Galveston Salt Marshes
Jones State Forest
Kempner Park
Sam Houston National Forest
Stubblefield Lake Recreation Area

Region III

Brazosport
Bryan Beach State Recreation Area
Davy Crockett National Forest
Spring Bluebonnet Trails

Region IV

Angelina–Neches Scientific Area
J.D. Murphree Wildlife Management Area
Pleasure Island
Roy E. Larsen Sandyland Sanctuary
Sea Rim State Park
Sidney Island

Region V

Aransas National Wildlife Refuge
North Toledo Bend Waterfowl Area
Old River Woodlands Trail
Sabine National Forest
Sylvan Nature Trail
Wild Azalea Canyon Trail

hours since the schedule varies. The museums are open Tuesday–Saturday, 9:00 A.M.–5:00 P.M. and Sunday–Monday, noon–5:00 P.M.

Lyndon B. Johnson Space Center
2101 NASA Rd. One, Houston, TX 77058—713-483-4321

Although not a nature outing as such, much can be learned about the moon through the exhibits at the visitor center here. Short briefings and self-guided tours are available daily from 9:00 A.M. to 4:00 P.M. The center is 25 miles southeast of Houston off Interstate 45.

Outdoor Activities

Aline McAshan Arboretum and Botanical Gardens
Memorial Park, 4501 Woodway, Houston, TX 77024—713-681-8433

This is also known as Houston Arboretum. More than five miles of hiking trails crisscross the 155 acres of native woodlands that sit on the banks of Buffalo Bayou. The park appears as most of the land surrounding Houston did before homes and freeways spread out from the center of town. Guided tours and lectures are given year-round.

Anahuac National Wildlife Refuge
P.O. Box 278, Anahuac, TX 77514—409-267-3131

More than 250 species of migratory birds stop off at this 10,021-acre refuge during the fall and winter seasons. Plenty of American alligators can also be seen here. The refuge is 20 miles south of Anahuac, Texas.

Angelina–Neches Scientific Area
District Office, Texas Department of Parks and Wildlife, Jasper, TX 75951—409-384-4335

Located west of Jasper on U.S. Route 190, the Texas Parks and Wildlife Department manages this 4,042-acre reserve on contract with the U.S. Corps of Engineers. Access to this area is at the juncture of the Neches and Angelina Rivers above Steinhagen Lake. It is by boat only, and there is no development. The area is used primarily by graduate students and serious nature lovers. There are no trails, but visitors can hike throughout the area.

Aransas National Wildlife Refuge
P.O. Box 100, Austwell, TX 77950—512-286-3559

This refuge is near Austwell, Texas, off State Route 35 on San Antonio Bay. The 54,829-acre preserve is home to over 350 species of birds in the winter, and many mammals, such as deer, javelinas, and raccoons, year-round. The refuge has played a major role in the return of the whooping crane from a low of 15 adults during the 1930s to over 150 in recent years.

Many well-marked nature trails and a platform with mounted telescopes offer visitors opportunities to view the waterscape and grassy islands of the refuge. Be sure you wear sturdy shoes, bring plenty of insect repellent, and watch for rattlesnakes.

Armand Bayou Nature Center
8600 Bay Area Blvd., Houston, TX 77058—713-474-2557

Raccoons, opossums, armadillos, deer, and wading birds are all very much in evidence at this 2,000-acre nature preserve in bayou country. Friends of the late naturalist Armand Yramategui purchased the site in his honor. The preserve contains prairie, estuary, and coastal woods. A greenhouse and nursery feature exhibits of native plants of the Texas gulf coast. Hiking trails crisscross the preserve, and free guided hikes and talks are given on the weekends. There are also "Owl Prowls" at 7:00 P.M. on first and third Wednesday nights from March through November. Visitors can also explore the estuary by pontoon boat or canoe. The preserve is open daily from 9:00 A.M. to 5:00 P.M.

Bay Area Park
c/o Clear Lake Park, 5001 NASA Rd. One, Seabrook, TX 77586—713-474-4891

This park is located on Armand Bayou. Its three distinct ecosystems—salt marsh, tall Gulf Coastal Prairie, and southern mixed hardwood forest—make this urban wilderness a naturalist's heaven. Bird watching is excellent. A 1,000-yard estuary walkway into the marsh gives visitors chances to view marsh wildlife. The park is open daily from 7:30 A.M. to 10:00 P.M.

Big Creek Recreation Area—Sam Houston National Forest
U.S. Forest Service, P.O. Drawer 1000, New Waverly, TX 77358—409-344-6205

This recreation area is only for hikers. Trails lead through a scenic area. Since there are no improvements, hikers must carry their own water. A five-mile trail leads to the Double Lake Recreation Area. This outing is ten miles outside Coldspring off State Route 150.

Brazos Bend State Park
Route 1, P.O. Box 840, Needville, TX 77461—409-553-3243

This 4,897-acre park extends along more than three miles of the Brazos River. It is located on Farm Road 762 about 20 miles southeast of Richmond. Both bird and wildlife are abundant here. Over 15 miles of trails, including marked nature trails and the three-mile Brazos River Trail are featured. An observation tower provides excellent views of the surrounding marshland, and wildlife observation and photography programs give visitors opportunities to explore the area. In 1989 the Houston Museum of Natural Science opened an observatory with a 36-inch research telescope at the park. The public is allowed to use the telescope on a first-come, first-served basis. Call the park for viewing schedules, which vary seasonally.

Brazosport

The beaches at Bryan, Quintana, and Surfside in the Brazosport area are all known nationally as stopover points for migratory birds. During the winter, bird watching is excellent here.

Bryan Beach State Recreation Area
c/o Park Superintendent, P.O. Box 1066, Sabine Pass, TX 77655—409-971-2559

About five miles outside Freeport, this flat park has a large inland tidal slough that is an ideal place for fishing and bird watching. The 878 acres in this preserve are bordered by the Gulf of Mexico, the Intracoastal Waterway, and the Brazos River.

Davy Crockett National Forest
U.S. Forest Service, 1240 E. Loop 304, Crockett, TX 75835—409-544-2046

Recreation areas such as Ratcliff Lake, Neches Bluff, and Kickapoo are located in this 161,478-acre reserve in Houston and Trinity Counties. The headquarters are in Crockett.

Double Lake Recreation Area—Sam Houston National Forest
U.S. Forest Service, P.O. Drawer 1000, New Waverly, TX 77358—409-344-6205

See the entry for Big Creek Recreation Area.

Edith L. Moore Nature Sanctuary
440 Wilchester Blvd., Houston, TX 77079—713-932-1392

This is a 17-acre woodland preserve run by the Houston Audubon Society. Nature trails border Rummel Creek and offer good birding. The sanctuary is open daily from dawn to dusk.

Galveston Island State Park
Route 1, P.O. Box 156A, Galveston, TX 77551—409-737-1222

This 2,000-acre park features nature trails and observation towers. It is bordered by the Gulf of Mexico and beaches to the south; and Galveston Bay and salt marshes to the north. Bird watching is spectacular here.

Galveston Salt Marshes
c/o Galveston Convention and Visitors Bureau, Galveston, TX 77550—713-763-4311

This is one of the best bird-watching sites in Galveston.

Houston Garden Center
1500 Hermann Dr., Houston, TX 77004—713-539-5371

In this garden, camellias often bloom as early as January, azaleas in March, and over 3,000 roses in April. There is little other nature here, but birds and butterflies are attracted to the many flowers.

Houston Hiking Trails

A network of trails join Houston-area parks from Memorial Drive to Allen's Landing. For a map of these trails contact either the Houston Parks and Recreation Department, P.O. Box 1562, Houston, TX 77001; 713-641-5051, or Citizens Environmental Coalition, Suite 1016, Main Plaza, Houston, TX 77002; 713-228-0037.

J.D. Murphree Wildlife Management Area
10 Parks and Wildlife Dr., Port Arthur, TX 77640—409-736-2551

About five miles west of Port Arthur off State Route 73, this preserve is accessible only by boat. These may be rented at nearby Myrtyle's Landing. The area offers excellent opportunities for bird watching during the fall migration of waterfowl, which use the Great Central Flyway. There is also a dense alligator population.

Jones State Forest
Route 7, P.O. Box 151, Conroe, TX 77301—409-273-2261

This 1,725-acre forest was logged in 1892, burned in 1923, and made a state forest in 1926. There are hiking trails throughout the forest, as well as the self-guided Sweetleaf Nature Trail in the northwest corner. The forest is six miles southwest of Conroe.

Kempner Park
Avenue O at 27th, Galveston, TX 77550

This is a city park that offers good bird watching during migration times. Palms, live oaks, and many flowering shrubs make this an excellent place to picnic.

Memorial Park
c/o Houston Parks and Recreation Department, P.O. Box 1562, Houston, TX 77001—713-641-5051

This 1,500-acre park has running, hiking, and biking trails, in addition to a wilderness area. The Houston Arboretum and Nature Center is located here. Memorial Park is bounded on the west by the Loop of Interstate 610. The park's northern boundary is Interstate 10, and the southern boundary is Buffalo Bayou.

Mercer Park Arboretum
22306 Aldine–Westfield Rd., Humble, TX 77338—713-443-8731

This 15-acre preserve has hiking trails in natural woodlands and garden areas. It is open daily from 8:00 A.M. to 5:00 P.M.

North Toledo Bend Waterfowl Area
Joaquin, TX 75954—409-598-5601

This 3,600-acre waterfowl area is being developed as a feeding ground for ducks. It offers good bird watching. The area is about seven miles southeast of Joaquin.

Old River Woodlands Trail
District Office, Texas Department of Parks and Wildlife, Jasper, TX 75951—409-384-4335

Follow U.S. Route 190, 12 miles west of Jasper. More than 50 species of trees, shrubs, and vines are identified by signs along this abandoned railroad right-of-way. It leads into the swamps of the Angelina River bottomland. Many of these species are rare or unusual. This Texas Forestry Association trail is accessible only by boat across Steinhagen Lake.

Pleasure Island

P.O. Box 1089, Port Arthur, TX 77640—409-983-3321

Near Port Arthur off State Route 82, this is a 3,500-acre island park in Lake Sabine. It is a great place to observe ships on the Intracoastal Waterway and sailboats on Lake Sabine. Nature trails lead visitors near hundreds of bird species, raccoons, opossums, and other small mammals. On the island, there is 24-hour crabbing, fishing along 16 miles of lakefront, and the world's longest fishing pier. The park is open daily from 5:00 A.M. to 10:00 P.M.

Roy E. Larsen Sandyland Sanctuary

c/o Texas Nature Conservancy, Silsbee, TX 77656—409-385-4135

This 2,200-acre preserve is under the protection of the Texas Nature Conservancy. More than six miles of trails wind through the pine and hardwood forests that border Village Creek. The Conservancy provides guided hikes by reservation, and the winding creek offers an excellent one-day canoe trip. The sanctuary is west of Silsbee on State Route 327

Sabine National Forest

101 S. Bolivar, San Augustine, TX 75972—409-275-2632

Located off State Route 87 near Shelbyville, this is the largest national forest in Texas, with 189,451 acres. It offers many scenic and recreational opportunities. The largest lake in Texas, the Toledo Bend Reservoir, has 650 miles of shoreline and 181,600 surface acres. It is bordered by the forest. There are a number of service roads leading into the forest, and six recreation areas maintained by the U.S. Forest Service.

Sam Houston National Forest

P.O. Drawer 1000, New Waverly, TX 77358—409-344-6205

One of four national forests in Texas, this one has 160,000 acres in three counties—Montgomery, Walker, and San Jacinto. It includes six major recreational areas and the 100-mile Lone Star Hiking Trail. A 26-mile segment of the trail begins just west of Cleveland, runs from the western tip of the Sam Houston National Forest near Richards, northeast toward Huntsville and then southeast toward Cleveland. This trail was built by the Lone Star Chapter of the Sierra Club.

Sea Rim State Park

P.O. Box 1066, Sabine Pass, TX 77655—409-971-2559

Off State Route 87 about 25 miles south of Port Arthur, this is the only marshland park in Texas. The 15,109 acres offer visitors excellent bird watching, photography, and over five miles of beaches. Waterfowl and small mammals such as muskrats and raccoons, as well as an occasional endangered red wolf, can be seen from canoe and airboat. Good views are also offered from the 3,640-foot boardwalk called the Gambusia Trail.

Sidney Island

From March through July this 17-mile island on the west bank of Lake Sabine is covered with nesting birds, including the colorful roseate spoonbill. The island is an Audubon Society sanctuary, and is not accessible to the public. But, bird-watchers can rent boats at Pleasure Island (see previous entry) and take a trip around Sidney Island for a good view of the nesting birds.

Spring Bluebonnet Trails

c/o Washington County Chamber of Commerce, 314 S. Austin, Brenham, TX 77833—409-836-3695

Through a cooperative effort including local townspeople, the Texas Department of Highways, Texas A&M University, and mother nature, the blue and white blossoms of the Texas state flower carpet the land in the spring. This display generally occurs during April, but has appeared as early as mid-March. Visitors can receive updates on the wildflowers by calling the Washington County Chamber of Commerce. They also provide maps of the most extensive blooms each year.

Stubblefield Lake Recreation Area

c/o Sam Houston National Forest, P.O. Drawer 1000, New Waverly, TX 77358—409-344-6205

Hiking trails, as well as camping facilities, are in this area located on the West Fork of the San Jacinto River. The area is about 15 miles west of New Waverly.

Sylvan Nature Trail

c/o Texas Forest Service, Route 7, P.O. Box 151, Conroe, TX 77301—409-273-2261

One of 15 Texas Woodland Trails sponsored by the Texas Forestry Association, this hiking path winds for over a mile through a varied forest of trees and shrubs, including numerous dogwoods. There is a roadside park opposite the trail head off U.S. Route 190 four miles southeast of Newton.

Wild Azalea Canyon Trail

c/o Texas Forest Service, Route 7, P.O. Box 151, Conroe, TX 77301—409-273-2261

Another of the 15 Texas Woodland Trails sponsored by the Texas Forestry Association, this one is most popular in March when the azaleas are in full bloom. Many other flowering plants can be found in this uncultivated forest during most of the spring. It is about 15 miles northeast of Newton off State Route 87.

Organizations That Lead Outings

Armand Bayou Nature Center Tour
8600 Bay Area Blvd., Houston, TX 77058—713-474-2551

Captain Brow Tours
Sea Gun Sports Inn, Route 1, P.O. Box 85, Rockport, TX 78382—512-729-2341
Boat tours of the Aransas National Wildlife Refuge are available to view whooping cranes. Tours run from October 20 to April 10. From five to 30 cranes are seen on each four-hour trip.

A Special Outing

When pioneers came to east Texas they encountered a thicket many believed to be almost 3,000 square miles. While this was an exaggeration, groups of thickets that were nearly impenetrable did cover an area nearly that large. Only wild animals seemed to inhabit this region of hardwood forests and stands of virgin white pine, both of which had thick undergrowth.

Tens of thousands of acres of this primeval habitat fell victim to timber operators and oil companies during the first part of the 20th century. Today a little over 100,000 acres of this natural jumble, with its fire ants, armadillos, mosquitoes, wildflowers, birds, and occasional bear and wolf, survive. The National Park Service administers 84,000 acres as the Big Thicket National Preserve. An additional 30,000 acres is in private preserves.

Visitors to the region find it easy to forget that they are in part of a national park once they enter the groomed trails that traverse the thickets. The trails range from one to nine miles long. These lead through forests covered with Spanish moss, filled with birds, and carpeted with wildflowers.

When Congress set aside the 84,000 acres, many residents wondered why anyone would want to protect mosquito-infested swampland that was so inhospitable. Visitors, especially those who come in April when the temperatures are in the 70s and the mosquitoes have yet to spring from the stagnate swamp waters, have no doubts. Yellowwood sorrel, violets, and jack-in-the-pulpit burst with color beneath the magnolia, cypress, and birch.

In addition to the miles of trails that lead through the thickets, there are many miles of slow-moving streams and rivers where canoeists enjoy leisurely trips. The National Park Service lists over 50 miles of trails in their various guidebooks for the Big Thicket. These guides also include maps of the region, information about campgrounds, and lists of outfitters that lead tours in the area.

For more information contact the Big Thicket National Preserve, 3785 Milam, Beaumont, TX 77701; 409-839-2689.

Nature Information

Houston Parks and Recreation Department
2999 S. Wayside, Houston, TX 77028—713-641-4111

Texas Parks and Wildlife Department
4200 Smith School Rd., Austin, TX 78744—800-792-1112, 512-479-4800 in Austin

Tourism Division
Texas Department of Commerce, P.O. Box 12008, Austin, TX 78711—800-888-8839

U.S. Forest Service
701 N. First St., Lufkin, TX 75901—409-592-6462

Further Reading

Ajilvsgi, Geyata. *Wildflowers of Texas.* Bryan, TX: Shearer Publishing, 1986.

Bomar, George W. *Texas Weather.* Austin: University of Texas Press, 1983.

Davenport, John C. *Houston.* 4th Ed. Austin: Texas Monthly Press, 1985.

Little, Mickey. *Campers Guide to Texas Parks, Lakes, and Forests.* 2nd Ed. Houston: Gulf Publishing Co., 1983.

Miller, George, and Delena Tull. *Texas Parks and Campgrounds.* Austin: Texas Monthly Press, 1984.

Nolen, Ben, and R. E. Narramore. *Texas Rivers and Rapids: Canoe and Backpack Guidebook.* Pipe Creek, TX: Nolen and Narramore, 1986.

Peterson, Roger Tory. *A Field Guide to the Birds of Texas.* Boston: Houghton Mifflin, 1963.

Rafferty, Robert R. *Texas Coast.* Austin: Texas Monthly Press, 1986.

Schueler, Donald G. *Adventuring Along the Gulf of Mexico.* San Francisco: Sierra Club Books, 1986.

Sheldon, Robert. *Roadside Geology of Texas.* Missoula, MT: Mountain Press, 1979.

Tennant, Alan. *The Snakes of Texas.* Austin: Texas Monthly Press, 1980.

Vanderbloom, Gretchen. *Texas Parks Guide.* Wauwatosa, WI: Affordable Adventures, 1988.

Dallas

Dallas was built on commerce. First the great cattle trails converged on Dallas and neighboring Fort Worth, then the area became a regional banking center for north Texas and the inland cotton markets. These all made Dallas a prosperous, but sleepy town until the big oil strike in east Texas in 1930.

From that boom, Dallas and Fort Worth have grown to one of the top metropolitan areas in the United States. Dallas is the nation's seventh largest city, with skyscrapers and modern business and art centers. As early as 1908 congestion and flooding along the Trinity River forced the citizens of Dallas to develop proposals to solve these problems. Some proposals included plans for moving the city away from the river, but one planning engineer proposed straightening the river instead. Today the Trinity River is little more than an engineered canal that passes through the city.

ABOVE: Both exotic and native animals can be viewed in a natural setting at the Fossil Rim Wildlife Center. Photo by Jimmie Munger.

Climate and Weather Information

From November to March, Dallas has what would be considered a very long spring to many residents of the North and Northeast. Daytime highs average between 30° and 65°F. Some days are brisk and windy, but it rarely snows. Most days are bright and sunny. During April and October temperatures rise to the mid-70s for even more enjoyable weather for naturalists outings.

There is a long summer, from May through September, when temperatures climb into the high 90s, and outings can become more uncomfortable. You should expect rain during the spring and summer months.

Getting Around Dallas

The streets of Dallas are not laid out in a traditional grid pattern, and visitors should refer to a map when driving downtown. There is almost no on-street parking, but there are plenty of lots and garages with reasonable rates. Taxi rates and availability are average, and car rental agencies are plentiful. Public transportation is good; for those outings in downtown and suburban areas, it is the preferred way to go. Most of the outings listed, however, are away from the transit routes.

Indoor Activities

Dallas Civic Garden Center

Second & Parry Sts., Dallas, TX 75226—214-428-8351

This two-story building houses many tropical plants. Sloping paths, catwalks, and spiral staircases lead visitors through the exhibits and past waterfalls. The center is open Monday–Friday from 10:00 A.M. to 5:00 P.M., and Saturday–Sunday from 2:00 P.M. to 5:00 P.M.

Fort Worth Museum of Science and History

1501 Montgomery St., Fort Worth, TX 76107—817-732-1631

This museum began in two spare rooms of a Fort Worth elementary school in 1941, when two teachers wanted to share their science and history materials with others. The museum now averages over one million visitors a year in its modern facilities, which opened in 1983. Exhibits on rocks and fossils, a planetarium, and the Omni Theater with a domed screen 80 feet in diameter are all included. This facility is open Monday–Thursday, 9:00 A.M.–5:00 P.M., Friday–Saturday, 9:00 A.M.–8:30 P.M., and Sunday, noon–5:00 P.M.

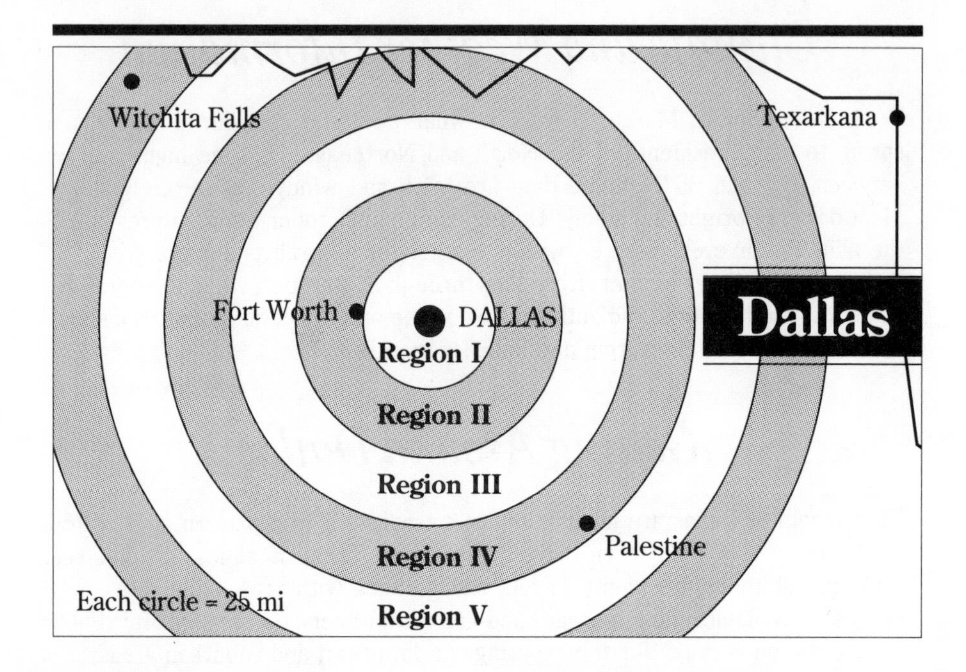

Indoor Activities

Region I
Dallas Civic Garden Center
State Fair Park

Region II
Fort Worth Museum of Science and History
Heard Natural Science Museum and Wildlife Sanctuary

Outdoor Activities

Region I
Dallas Arboretum and Botanical Gardens
Green Hill Environmental Center
L.B. Houston Nature Trail
Turtle Creek
White Rock Lake and Park
Woodland Basin Nature Area

Region II
Eagle Mountain State Fish Hatchery
Fort Worth Botanic Gardens
Fort Worth Nature Center and Refuge
Heard Natural Science Museum and Wildlife Sanctuary

Region III
Hagerman National Wildlife Refuge

Region IV
Davey Dogwood Park
Gambill Goose Refuge
Gus Engelking Wildlife Management Area

Region V
Atlanta State Recreation Area
Caddo Lake State Park

Heard Natural Science Museum and Wildlife Sanctuary
Farm Road 1378, McKinney, TX 75069—214-542-5566

Located off State Route 5 south of McKinney, this museum has a permanent collection of seashells, rocks, and minerals from north Texas. It also includes a human-made cave complete with stalactites and stalagmites. See the Outdoor Activities entry for more information. This facility is open Tuesday–Saturday, 9:00 A.M.–5:00 P.M. and Sunday, 1:00 P.M.–5:00 P.M.

State Fair Park
c/o Friends of Fair Park, P.O. Box 26248, Dallas, TX 75226—214-426-3400

This park is off Interstate 30 on U.S. Route 67/80. In addition to the state fair buildings and the Cotton Bowl, several museums and other nature exhibits are located here. Among them are the following.

Dallas Aquarium has over 4,000 fish, a giant octopus, a 35-pound Maine Lobster, and plenty of alligators. Call 214-428-3587 for more information. The Dallas Museum of Natural History includes a reconstructed skeleton of a mammoth whose bones were found near the Trinity River. Outstanding wildlife dioramas show animals and plants in their native Texas habitats. Call 214-421-2169 for more information.

The Southwest Museum of Science and Technology—Science Place II features permanent exhibits on energy and ecology, plus many other traveling exhibits. The Dallas Planetarium is located in Science Place II. Call 214-428-5555 for information. All these museums are open Monday–Saturday, 9:00 A.M.–5:00 P.M. and Sunday, noon–5:00 P.M.

Outdoor Activities

Atlanta State Recreation Area
Route 1, P.O. Box 116, Atlanta, TX 75551—214-796-6476

This 1,475-acre park is off U.S. Route 59 south of Texarkana. Lake Wright Patman, which is over five miles across at its widest point and 25 miles from end to end, is the main attraction. There are also several hiking trails that wander through the rolling woodlands and the bottomlands of Wilkins Creek. One trail is over three miles long. There is also a self-guided woodland nature trail.

Caddo Lake State Park
Route 2, P.O. Box 15, Karnack, TX 75661—214-679-3351

A visitor center, interpretive and hiking trails, and lake provide a variety of naturalist activities at this state park. It was originally built by the Civilian Conservation Corps during the 1930s.

Dallas Arboretum and Botanical Gardens
8617 Garland Rd., Dallas, TX 75218—214-327-8263

This historic 66-acre former estate overlooks White Rock Lake, and offers a pleasant walk among many plants. The gardens are open Tuesday–Sunday from 10:00 A.M. to 6:00 P.M.

Davey Dogwood Park
North Link Rd., Palestine, TX 75801—214-729-6066

Paved roads wind through this 400-acre park that features waterways and blooming dogwoods in the spring. One of the Texas Dogwood Trails is part of this park.

Eagle Mountain State Fish Hatchery
Eagle Mountain Circle, Azle, TX 76020—817-237-3536

From early April through mid-May visitors can observe the hatching process of bass and catfish, and learn about their life cycles. These fish are planted in Eagle Mountain Lake. The hatchery is on Eagle Mountain Circle off Ten Mile Bridge Road northwest of Fort Worth, Texas.

Fort Worth Botanic Gardens
3220 Botanic Garden Dr., Fort Worth, TX 76107—817-870-7685

Over 100 acres of native and exotic plants offer visitors a beautiful outing in the city. A Japanese garden is also featured. This facility is open Tuesday–Sunday from 9:00 A.M. to 7:00 P.M.

Fort Worth Nature Center and Refuge
Route 10, P.O. Box 53, Fort Worth, TX 76135—817-237-1111

This 3,400-acre refuge sits on land that was never developed. It is home to animals such as buffalo and prairie dogs. Self-guided trails lead through the Prairie Dog Town, to the top of Caprock, and through lotus marshes. Some trails are accessible to wheelchairs and strollers. The refuge is open Monday–Friday, 8:00 A.M.–5:00 P.M., and Saturday–Sunday, 9:00 A.M.–5:00 P.M.

Gambill Goose Refuge
Farm Road 2820, Paris, TX 75460—214-784-2501

About seven miles northwest of Paris, this state wildlife refuge was named in honor of John C. Gambill. He began feeding greater Canada geese on his farm in 1922 and continued during the next 35 years. The state now handles the job on 624 acres along the shore of Lake Gibbons.

Green Hill Environmental Center
7575 Wheatland Rd., Dallas, TX 75249—214-296-1955

Many rare plants and animals can be seen at this wilderness environmental center. With 1,000 acres, two ponds, and many trails, visitors enjoy this outstanding resource. Portions of the White Rock Escarpment are located within the center's land. Free guided hikes are given every weekend, with special tours of a 40-acre trail held on Saturday. Special programs are given during the summer. Call for information on current activities.

Gus Engelking Wildlife Management Area

Palestine, TX 75801—214-928-2251

Upland songbirds of the post-oak forests can be seen here in abundance, along with some waterfowl. Fishing is good on Catfish Creek. This area is on U.S. Route 287, about 20 miles northwest of Palestine.

Hagerman National Wildlife Refuge

Route 3, P.O. Box 123, Sherman, TX 75090—214-786-2826

Primarily a refuge for migrating waterfowl, ducks and geese are almost always seen. This site is also excellent for viewing migratory shorebirds during early spring and late summer. Sandpipers, plovers, and phalaropes are especially abundant. This refuge is about 15 miles northwest of Sherman off U.S. Route 82.

Heard Natural Science Museum and Wildlife Sanctuary

Farm Road 1378, McKinney, TX 75069—214-542-5566

More than 240 species of animal life are found at this sanctuary. Several can be seen on the nature trail. Although many of the animals are nocturnal, they can be seen in at rest during daylight hours. Injured animals that are being, or have been, rehabilitated can also be viewed as they are readied for return to the wild. A trail is accessible to wheelchairs and strollers. See Indoor Activities for more information.

L.B. Houston Nature Trail

Tom Braniff Dr., Dallas, TX 75235—214-421-2169

The Dallas Museum of Natural History maintains these nature trails near the Elm Fork of the Trinity River. Visitors can pick up a trail pamphlet at the museum. The trails are rough, and not always highly maintained, but do offer visitors opportunities to observe riparian life.

Turtle Creek

Turtle Creek Park, 3400 Turtle Creek, Dallas, TX 75219—214-369-8451

A number of city parks are located along this stretch of creek, which runs from Maple and Oaklawn Avenues, along Turtle Creek Parkway to Highland Park. This is one of the loveliest areas of Dallas. Reverchon, Lee, and Turtle Creek Parks all have nature and recreation facilities, and the areas along the creek offer a peaceful escape from the city.

White Rock Lake and Park

830 E. Lawther, Dallas, TX 75218—214-321-2125

This city lake has an eight-mile biking and hiking trail around it. While it is often crowded, the park does offer an escape for city visitors. Fossils can be found in the limestone ledges near the lake spillway. Birds such as grackles, cowbirds, starlings, and red-winged blackbirds inhabit the bamboo thicket near the lake. Can Dyke Hill is carpeted with wildflowers from spring through early fall.

Woodland Basin Nature Area
East Miller Rd., Garland, TX 75041—214-670-7070

This area is swamp and bottomland that is home to a wide variety of birds, reptiles, and small mammals. A boardwalk leads visitors through the lowlands, and offers a good view of reeds and other swamp plants. The best times to visit are spring, fall, or winter, for the muggy heat of mid-summer can be enervating.

Organizations That Lead Outings

The Dallas Nature Center
7575 Wheatland Rd., Dallas, TX 75249—214-296-1955

Dallas Safari Club
8585 Stemmons Freeway, Dallas, TX 75247—214-630-1453

Sierra Club—Southern Plains Office
6220 Gaston, Dallas, TX 75214—214-824-5930

A Special Outing

This may not be considered a true nature outing, for you never need to get out of your car during the nine-mile drive through the Fossil Rim Wildlife Center. This is an 1,800-acre haven for rare and endangered species.

Over 1,000 animals representing 24 species live within this internationally acclaimed program. From Grevy's zebra to cheetah, endangered animals are bred here. That gives visitors opportunities to view animals walking freely about the habitat.

The center is now introducing programs to help increase the population of such nearly extinct species as the White rhinoceros and Red wolf. The White rhinoceros prefer to live in groups, unlike the solitary pairs of Black rhino. Fossil Rim is attempting to duplicate the White rhino's living conditions and arrangements.

There are fewer than 100 Red wolves known to exist, and the species is highly endangered. Five mating pairs have been transported to Fossil Rim to form the nucleus of a breeding program. Resident biologists hope the animals can be reintroduced into the wild.

Visitors to the Fossil Rim Wildlife Center can observe many of these endangered animals roaming free in an environment much like their native ones. You can gain knowledge of the many programs being conducted at the center to help preserve endangered species.

The center is about a one-hour drive from Forth Worth, and is three miles west of Glen Rose, Texas. During the summer, it is open daily from 9:00 A.M. to one hour before sunset. The center is closed during the winter. For more information contact the Fossil Rim Wildlife Center, P.O. Drawer 329, Route 1, Box 210, Glen Rose, Texas; 817-897-2960.

Nature Information

Dallas Department of Parks and Recreation
City Hall, 1500 Merulla, Dallas, TX 75201—214-670-4100

Dallas Visitor Information Center
400 S. Houston St., Dallas, TX 75202—214-954-1111

Fort Worth Parks and Recreation Department
2212 Forest Park Blvd., Fort Worth, TX 76110—817-927-9386

Texas Forestry Association
P.O. Box 1488, Lufkin, TX 75901—409-632-TREE

Texas Parks and Wildlife Department
4200 Smith School Rd., Austin, TX 78744—800-792-1112,
512-479-4800 in Austin

Tourism Division
Texas Department of Commerce, P.O. Box 12008, Austin, TX 78711—
800-888-8839

U.S. Forest Service
701 North First St., Lufkin, TX 75901—409-592-6462

Further Reading

Also see the Houston *Further Reading* section for other books on Texas.

Jackson, Joan F., and Glenna Whitley. *Places to Go With Children in Dallas and Fort Worth.* San Francisco: Chronicle Books, 1987.

Sherley, Connie. *Texas and the Southwest.* New York: Fisher Travel Guides, 1986.

Denver

The Rocky Mountains, the most prominent feature of Colorado, rise west of Denver, and overshadow all that goes on in and around the city. East of the city, the Great Plains extend through the eastern third of Colorado, Kansas, and Nebraska.

As Denver grew from a dusty cowtown to a modern metropolitan area, it did not neglect its natural heritage. In addition to a series of city parks in the downtown area, the city preserved large sections of the nearby mountains in a unique park system. The Denver Mountain Park system is a series of 49 municipally owned parks outside the city that contain over 25,000 acres of mountain, woodland, and canyon scenery. These areas provide visitors and residents opportunities to wander in rugged Rocky Mountain parks that are undisturbed by human encroachment.

The western mountains consume much of the attention paid to naturalist outings near Denver. But, there are a variety of activities to the east along the rivers that run from the mountains to the plains.

ABOVE: There is plenty of bird life for these visitors to view at the Barr Lake Wildlife Refuge and Nature Center outside Denver. Photo courtesy of Colorado Division of Wildlife.

Climate and Weather Information

Denver is noted as the "Mile-high City," and its elevation exerts a moderating influence on the summer climate. From May through September visitors can usually expect warm, comfortable days, and cool evenings. There are occasional heat spells that take the temperature near 100°F. As you leave the city and head higher into the mountains the temperature drops.

Fall and spring are both likely to have crisp days intertwined with warmer periods and cold nights. Winter brings cold, snow, and wind to make it the most uncomfortable time for outdoor activities; although many sunny, clear, and crisp days are excellent for exploring nature during its winter hibernation. Spring usually brings rain, and there are some summer storms, especially at higher elevations.

Getting Around Denver

Denver has developed a good freeway system that allows visitors to easily go from one section of town to another. Downtown streets generally run in a grid pattern. On-street parking is limited, but lots and garages are plentiful with inexpensive rates. Rapid transit is only adequate, and will not take you to most of the activities listed, but car rental agencies are numerous. Taxis are not as readily available in Denver as in some other cities in this guide, but fares are reasonable.

Indoor Activities

There are few indoor nature activities in the Denver area. The climate is so conducive to outdoor activities, however, that visitors do not need many indoor activities to keep them busy.

Colorado School of Mines Geology Museum
16th & Maple Sts., Golden, CO 80401—303-273-3823

This museum features an extensive display of rocks, minerals, and fossils, and an early Colorado mine exhibit. It is open Monday–Saturday, 9:00 A.M.–4:00 P.M. and Sunday, 1:00 P.M.–4:00 P.M.

Denver Museum of Natural History
Montview & Colorado Blvds., Denver, CO 80207—303-322-7009

This is one of the largest natural history museums in the country and features both regional and worldwide exhibits. The museum is open daily from 9:00 A.M. to 5:00 P.M. It is in City Park, which is bounded by 17th and 23rd Avenues, York Street, and Colorado Boulevard.

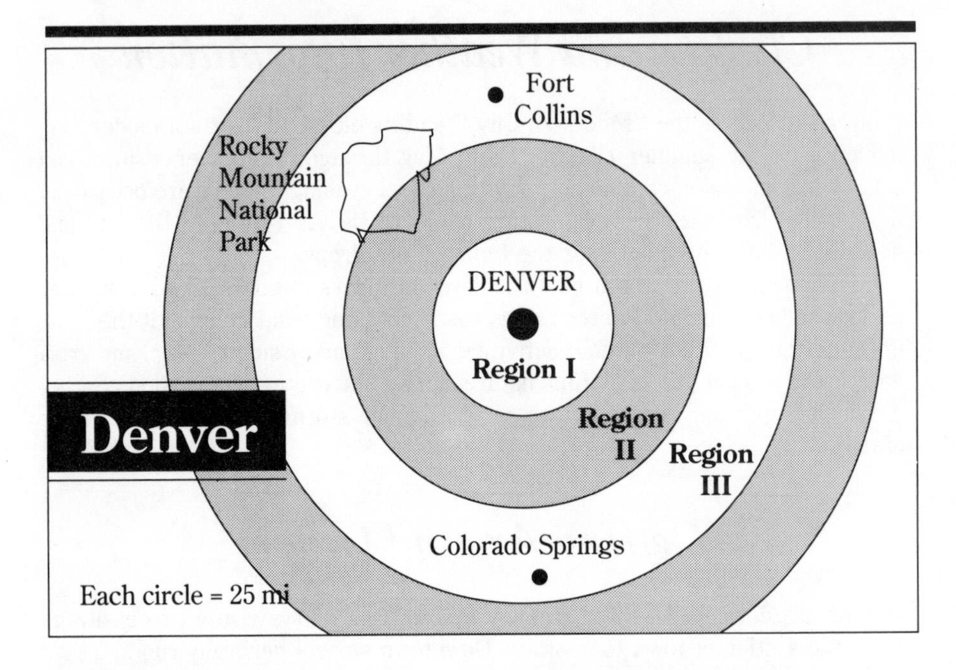

Indoor Activities

Region I

Colorado School of Mines Geology Museum
Denver Museum of Natural History
Jefferson County Conference and Nature Center

Region II

National Center for Atmospheric Research
University of Colorado Museum

Region III

Fort Collins Museum

Outdoor Activities

Region I

Denver Botanic Gardens
Jefferson County Conference and Nature Center
Mountain View Park
Peregrine Falcons Viewing Booth
Platte River Greenway
Sloan's Lake Park
Washington Park

Region II

Arapaho National Recreation Area
Barr Lake
Boulder Creek
Chatfield State Recreation Area
Cherry Creek State Park
Denver Mountain Park System
Eldorado Canyon State Park
Golden Gate Canyon State Park

Region III

Florissant Fossil Beds National Monument
Lory State Park
Rocky Mountain National Park

Fort Collins Museum
200 Mathews St., Fort Collins, CO 80524—303-221-673

This museum specializes in the human and natural history of northern Colorado. Although human history is dominant, there are some good natural history exhibits of the region. The museum is open Tuesday–Saturday, 10:00 A.M.–5:00 P.M. and Sunday, noon–5:00 P.M.

Jefferson County Conference and Nature Center
900 Colorow Rd., Golden, CO 80401—303-526-0855

This center features exhibits of the ecosystem and wildlife of the Front Range. From April to October, it is open Tuesday, Thursday, Saturday, and Sunday from 10:00 A.M. to 4:00 P.M. From November to April, the hours are the same; however the center is closed on Sundays. See entry in Outdoor Activities for more information.

National Center for Atmospheric Research
1850 Table Mesa Dr., Boulder, CO 80303—303-497-1174

Self-guiding tours of exhibits on solar astronomy, weather and various atmospheric phenomena are featured. The center is open Monday–Friday from 8:00 A.M. to 5:00 P.M.

University of Colorado Museum
Henderson Bldg., University of Colorado, Boulder, CO 80302—303-492-6892

This museum has many artifacts and exhibits on the natural history of Colorado. It is open Monday–Friday, 9:00 A.M.–5:00 P.M., Saturday, 9:00 A.M.–4:00 P.M., and Sunday, 10:00 A.M.–4:00 P.M.

Outdoor Activities

Denver is an outdoor city. Its naturalist activities vary from those that take visitors to the east onto the plains, to those that reach over 14,000 feet high at the top of the Rockies. This elevation range allows visitors to partake of activities that, in other parts of the country, can only be undertaken over several seasons.

Arapaho National Recreation Area
U.S. Forest Service, 11177 W. 8th Ave., Lakewood, CO 80225—303-887-3331

Adjacent to the southern edge of the Rocky Mountain National Park off U.S. Route 40, this 36,000-acre area includes many lakes and trails. Between December and May, the lakes are frozen, and most of the trails are covered with snow. The best time to visit the area is June through August, when the weather is most predictable.

Barr Lake
c/o Colorado Division of Parks and Outdoor Recreation, 1313 Sherman, Denver, CO 80203—303-659-6005

Northeast of Denver off Interstate 76, this 2,609-acre state park has a lake, nature center, and trails. There is cross-country skiing on the trails during the winter.

Bike Paths

A map of the extensive bike paths of Denver, which can also be used for hiking, is available. Call or write the Bicycle Racing Association of Colorado, 1290 Williams Street, Denver, CO 80218; 303-333-2453.

Boulder Creek

A four and one-half-mile stretch of this creek that runs through the heart of Boulder has been restored through a three-year project that involved businesspeople, political leaders, and environmentalists. There is an uninterrupted biking and hiking path along the creek, which is open to intertubing, rafting, and kayaking most of the year.

Native vegetation that has survived years of punishment has been restored through clearing and new planting. With this work the creek lies in an almost natural state.

Chatfield State Recreation Area

11500 N. Roxborough Park Rd., Littleton, CO 80125—303-797-3986

From April to September visitors can observe herons nesting in tops of dead trees in the middle of the reservoir. There is an overlook about 100 yards from the rookery, and visitors gain excellent views with spotting scopes or binoculars. A visitor's center and nature trail are also located in this 6,750-acre park which is southwest of Denver off U.S. Route 85.

Cherry Creek State Park

c/o Colorado Division of Parks and Outdoor Recreation, 1313 Sherman, Denver, CO 80203—303-690-1166

Southeast of Denver on State Route 83, this is a multi-purpose state park on 4,795 acres. Many activities, including a shooting range, attract large crowds, but nature trails lead naturalists away from the noise of shooters and water-skiers.

Denver Botanic Gardens

1005 York St., Denver, CO 80206—303-575-2547

Both the outdoor gardens and a conservatory have native and exotic plants in beautifully landscaped arrangements. The conservatory has over 800 plant species. It is open daily from 9:00 A.M. to 5:00 P.M.

Denver Mountain Park System

This park system is unique. With 49 municipally owned parks covering over 25,000 acres in the foothills of the Rockies, it offers great opportunities to visit some very scenic areas. The parks extend some 50 miles west of Denver and about 20 miles south. The nearest park is 13 miles from downtown.

A popular outing is the Denver Mountain Park Circle Drive that passes through Golden, up Lookout Mountain, and south past Genesee Peak, which is 8,270 feet high. It continues through Fillius and Bergen Parks before entering Bear Creek Canyon and Red Rock Park. From the town of Morrison, you take U.S. Route 6 north back to Denver. For more information on the parks and activities in this system, contact the Denver Department of Parks, 945 South Huron, Denver, CO 80223; 303-575-2552.

Eldorado Canyon State Park

c/o Colorado Division of Parks and Outdoor Recreation, 1313 Sherman, Denver, CO 80203—303-866-3437

Southwest of Boulder on State Route 93, this is a small, 845-acre park with good hiking trails. It is generally not very crowded, because the activities are limited to picnicking and hiking.

Florissant Fossil Beds National Monument

P.O. Box 185, Florissant, CO 80816—719-748-3253

This 6,000-acre monument protects Oligocene fossils that were preserved in paper-thin strata as layers of volcanic ash accumulated over an ancient lake bed. A visitors center, interpretive programs, and guided walks are offered here. Hiking trails lead into the backcountry. The visitors center is open daily from 8:00 A.M. to 5:00 P.M. This park is west of Colorado Springs off State Route 67.

Golden Gate Canyon State Park

Route 6, Box 280, Golden, CO 80403—303-592-1502

This park is northwest of Denver off State Route 119. Thirteen different trails, with over 40 marked miles, lead visitors along valley floors, over ridges, and to peaks with panoramic vistas. The trails vary in length from just over a mile to nearly six miles, and go from easy to strenuous.

Quiet forests, rushing streams, and grand vistas are all present in the spruce- and pine-covered slopes where deer, bears, elk, bobcats, and mountain lions live. Golden eagles and sharp-shinned hawks often soar overhead.

July and August are the best time of year for camping and hiking in the park, but its proximity to Denver means heavy weekend crowds. Late spring and early fall also offer excellent hiking weather, but it is more unpredictable than mid-summer.

Jefferson County Conference and Nature Center

900 Colorow Rd., Golden, CO 80401—303-526-0855

This center sits atop the summit of 7,500-foot Lookout Mountain, and offers a variety of activities. Inside the nature center, which is in an old, Tudor-styled carriage house, there are exhibits about the Front Range ecosystem and wildlife. Naturalists are available to answer questions.

Outside there is a one-and-one-half-mile loop trail that winds through a conifer forest onto a sloping meadow. To help visitors explore the trail on their own, the nature center provides a family adventure pack with wind gauge, compass, thermometer, magnifying glasses, and guide book.

Other trails lead through the 110 acres of this retreat, which was once an estate. Trails stay open until the first snow, usually around Thanksgiving. From April through October, the nature center is open Tuesday, Thursday, Saturday, and Sunday, from 10:00 A.M. to 4:00 P.M. From November to April, it is open the same hours, except that it is closed on Sundays.

Lory State Park
708 Lodgepole Dr., Bellevue, CO 80512—303-493-1623

This 2,600-acre park is located in Rocky Mountain high country, off State Route 287 near Bellevue. The elevation ranges between 5,000 and 7,000 feet. Over 25 miles of trails lead through a variety of vegetation. The park includes sagebrush and grassland valleys, sandstone hogback ridges, slopes covered with low-lying mountain shrubs, and ponderosa pine forests on higher ridges.

This park is a former cattle ranch. Wild turkeys, blue grouse, mule deer, coyotes, and the unusual Abert squirrels, which have long, tufted ears, roam the countryside.

Arthur's Rock, which rises to 6,780 feet, is a favorite rock climber's site. A trail also leads to the top of this peak, where there is a breathtaking view of Fort Collins, Colorado, and its surrounding areas.

Mountain View Park
East 3rd Ave. & Belaire St., Denver, CO 80220—303-575-2552

This park provides a good view of the mountains on a clear day, and a labeled profile of the Continental Divide in terrazzo tile. This is provided for visitors to use in identifying various landmarks.

Peregrine Falcons Viewing Booth
The Colorado State Division of Wildlife, 16th St. & Broadway, Denver, CO 80290—303-830-2557

Peregrine falcons once nested up and down the Front Range of the Rockies, but disappeared by the late 1970s. Due to a number of recovery projects over two dozen pair now nest in Colorado. In the summer of 1988, young peregrines were placed in nest boxes on top of a downtown Denver high rise. Similar projects have been undertaken in other cities with varying success, and there is hope that Denver's project will be successful.

The Colorado State Division of Wildlife has set up a viewing booth across the street from One Civic Center Plaza to view the birds as they hunt pigeons in the city sky. Call for information on the progress of the project, and when peregrines may be viewed.

Platte River Greenway

This green way runs for over ten miles along the Platte River through Denver. Its route leads through 17 parks and preserves as it goes from Interstate 70 on the north to U.S. Route 285 on the south. Although the green way is a narrow strip along the river amidst roads and heavy development, its four riverside trails and variety of parks and preserves give visitors and residents opportunities for a speedy escape from downtown pressures.

A children's museum is located in Gates-Crescent Park near the north end of the green way. The Overland Pond Educational Park near the south end includes a one-acre fly-casting pond and over six landscaped acres with plants from five Colorado ecological zones.

All in all, this green way belt is one of the best in the country. It offers an urban escape for all who are interested in nature.

Rocky Mountain National Park
Estes Park, CO 80517-8397—303-586-2371

Rocky Mountain National Park is one of the nation's most beautiful parks, and includes over 250,000 acres of the Rocky Mountains' Front Range. The region, with 68 named peaks that reach over 12,000 feet, is one of the highest in the contiguous states. Valleys are about 8,000 feet in elevation. The results of early glacial activity is clear to even the untrained eye. Both the wildflowers and wildlife are spectacular. Several museums and visitor centers in the park explain the region's natural history, and offer visitors a chance to explore the park in depth.

A fairly direct route leads from Denver through Boulder and Estes Park to the east side of Rocky Mountain National Park. There visitors enjoy the Moraine Park Museum and several outstanding outdoor areas.

For those who want to enjoy a beautiful drive there is a 240-mile loop from Denver through Boulder, Estes Park, Grand Lake, and Idaho Springs. It is one of the most impressive circle trips in the United States. This drive cuts across the Continental Divide on the Trail Ridge Road, one of the highest paved roads in the country. It then goes to Grand Lake, and returns to Denver via Berthoud Pass and through the Denver Mountain Parks.

Sloan's Lake Park
West 26th & Sheridan, Denver, CO 80214—303-575-2552

It takes about an hour to circle the lake on the shoreline foot trails. On clear days the view of the mountains is spectacular. Bird watching is a favorite pastime of many visitors here.

Washington Park
South Downing & East Virginia, Denver, CO 80209—303-575-2552

Small creeks, with trails alongside, flow under willows and cottonwoods here. Crowds from nearby Denver University spill into the park on nice days. There is not much true nature here, but the two lakes, flower gardens, and birds provide a pleasant break.

Organizations That Lead Outings

American Wilderness Alliance
7600 E. Arapahoe Rd., Englewood, CO 80112—303-771-0380

Colorado Mountain Club
2530 W. Alameda, Denver, CO 80228—303-922-8315

Denver Audubon Society
975 Grant, Denver, CO 80203—303-860-1471

Foothills Nature Center
4201 Broadway, Boulder, CO 80302—303-442-4288

Greenpeace
1310 College Ave., Boulder, CO 80302—303-440-3381

Nature Conservancy
1244 Pine, Boulder, CO 80302—303-444-2950

Sierra Club—Rocky Mountain Chapter
777 Grant, Denver, CO 80203—303-861-8819

Volunteer for Outdoors Colorado
1410 Grant, Denver, CO 80203—303-830-7792

Wilderness Society
777 Grant, Denver, CO 80203—303-839-1175

A Special Outing

This is not so much an account of a special outing, but of a special time of year. Most cities have regular seasons, each with pretty much the same pattern of time and duration. The length of these seasons vary with the cities. Those in the North and Northeast, for example, have short spring and fall seasons, with moderate summers and long winters. Those in the South have a short winter season, and longer spring, summer, and fall ones. Visitors to Denver discover another type of season. While Denver itself has a reasonable spring and fall, residents can claim to have prolonged seasons because of their geography.

By April, spring may come to Denver, and the prairie to the east. The tops of willows and cottonwood, along the streams where the elevation is from 3,500 to 5,500 feet, are tinted with new green growth. Local birds have begun building their nests. Horned larks are involved in their courtship activities, and Great Horned owls already have white, down-covered young demanding to be fed.

But spring is still a long way away in the higher elevations of the Rockies to the west. There, winter still has a grip on the ice- and snow-covered streams, and buds will not appear on stream-side vegetation for another two months. To make the journey from the prairie east of Denver, to the mountain tops west of the city, is equivalent to traveling northward through Canada to where the trees become dwarfed and windblown as they give way to the arctic tundra.

The change of seasons occurs slowly as you go from the prairie to the peaks. It begins in April along the prairie creek banks as buds green the tips of willow branches. It continues westward until roses bloom and birds reach the peak of their nesting season in Denver around the first of June. The high mountains are still likely to be covered with winter snow in June. Spring creeps up into the Transition Zone where flowers begin to dot the mountain slopes by mid-June. By early July, it brings showers to the mountain meadows and lakes of the Canadian Zone, along with masses of wildflower blooms. Finishing in mid-July and early August, alpine meadows burst into colorful carpets of wildflower blooms.

In no other city in this guide can one season be so long. Denver offers naturalists, particularly those interested in birds and wildflowers, unique opportunities to explore the progression of spring through all the life zones of the West with the exception of Upper and Lower Sonoran. To do so you simply travel several hundred miles from east to west.

Nature Information

Colorado Division of Wildlife
6060 Broadway, Denver, CO 80216—303-825-1192

Colorado Parks and Outdoor Recreation
1313 Sherman St., Denver, CO 80216—303-866-3437

Denver Parks and Recreation Department
945 S. Huron, Denver, CO 80223—303-698-4900

Denver Visitors Bureau
225 W. Colfax St., Denver, CO 80202—303-892-1112

Division of Commerce and Development
State Office of Tourism, 1313 Sherman St., Denver, CO 80203—303-592-5410

U.S. Forest Service
Rocky Mountain Region, 11177 W. 8th Ave., Lakewood, CO 80225—303-234-4368

Further Reading

Agnew, Jeremy. *Exploring the Colorado High Country.* Colorado Springs, CO: Wildwood Press, 1977.

Ayer, Eleanor. ed. *Colorado Traveler—Parks and Monuments: A Scenic Guide to Colorado.* Washington, DC: R H Publications, 1987.

Boddie, Peter, and Caryn Boddie. *The Hikers Guide to Colorado.* Billings, MT: Falcon Press, 1984.

Chronic, Halka. *Roadside Geology of Colorado.* Missoula, MT: Mountain Press, 1980.

Collins, Donna. *Colorado Traveler—Natural Sites: A Guide to Colorado's Natural Wonders.* Washington, DC: R H Publications, 1987.

Fielder, John. *Colorado Wildflowers Littlebook.* Englewood, CO: Westcliff Publishers, 1985.

Folzenlogen, Robert. *Birding Guide to Denver–Boulder Region.* Boulder, CO: Pruett Publishing Co., 1986.

Hagen, Mary. *Hiking Trails of Northern Colorado.* Boulder, CO: Pruett Publishing Co., 1983.

LeRoy, L.W., and D.A. LeRoy. *Red Rocks Park: Geology and Flowers.* Golden, CO: Colorado School of Mines, 1978.

McKinney, Lee, and D.T. McKinney. *Colorado Traveler—Gems and Minerals: A Guide to Colorado's Native Gemstones.* Washington, DC: R H Publications, 1987.

Perry, John, and Jane G. Perry. *The Sierra Club Guide to the Natural Areas of Colorado and Utah.* San Francisco: Sierra Club Books, 1985.

Rigby, J. Keith. *Field Guide: North Colorado Plateau.* Dubuque, IA: Kendall Hunt, 1975.

Phoenix

Phoenix, and its sister city Tucson, which lies some 100 miles to the southeast, are located in the largest section of true Sonoran desert in the United States. Visitors to the region are often overwhelmed by the broad, seemingly endless stretches of cactus and mesquite that surround the cities. Almost all naturalist activities are directly related to the desert. Visitors must be aware of the constrictions placed upon them by this sometimes strange and unfamiliar setting.

Deserts are hot and dry. These characteristics can scare first-time visitors, for the tales of people dying from heat and thirst after being lost in the desert have long been a staple in movies, cartoons, and books. These tales are based on reality, for daytime temperatures in the unshaded desert can exceed 120°F. There is little water to be found. These factors notwithstanding, the region does offer outstanding naturalist opportunities that can be enjoyed with a minimum of danger and discomfort.

ABOVE: Visitors flock to the Arizona-Sonora Desert Museum outside Tucson, where native Sonora Desert flora and fauna can be observed close-up. Photo by Tim Fuller.

The broad valleys around Phoenix and Tucson, and the fault-block mountains that separate them, are part of the basin and range province that covers Nevada, Utah, and Arizona. These areas have long been inhabited by people, but it was only with the advent of modern developments, such as air conditioning and large-scale irrigation and water storage projects, that the population boomed.

These developments made it possible for the region to maintain a large population year-round, and brought manufacturing plants that employed numerous people. These new residents have concentrated near the cities, rather than spread out over large sections of vacant land, for the desert is not the most hospitable place for individual dwellings.

This has left much of the land outside the fertile Gila and Salt River valleys in its natural state, and undisturbed by humans except for the occasional miner. Today's visitors who wish to experience the desert in its most primal state can make that wish come true.

Climate and Weather Information

To visitors from the North and East the climate of southern Arizona seems unbelievable. The state has over 300 days of sunshine a year, and less than seven inches of rainfall. This makes for cloudless days, and winter temperatures that are comparable to those of eastern springs.

It is not a perfect climate, though, for summer highs frequently exceed 100°F. Although this is generally a dry heat, since more moisture is lost through evaporation each year than is gained by rainfall, 115°F is hot and enervating to almost everyone. Also, summer is the rainy season when the remnants of monsoons from the Gulf of California bring clouds and rain during days with temperatures over 100°F. Therefore, you get some afternoons that approach a sauna in heat and humidity.

From fall through spring, however, the weather is almost always pleasant, and conducive to exploring the outdoors. During this time even the most timid can explore the desert without fear. This does not mean, however, that you should not take normal precautions to ensure that you do not get lost, for the desert is not a hospitable place, even in the winter.

Getting Around Phoenix

Both Phoenix and Tucson are relatively new cities that were built after the automobile became the dominant force of American life. Both their street layouts and freeway systems reflect that. The streets are laid out in a grid pattern, and are easy to navigate. Freeways serve most major areas of both cities.

Parking is plentiful, at least in comparison with eastern cities, and there are plenty of lots and garages with inexpensive parking rates. Public transportation is not as efficient as in older cities, and taxis are mainly available in the central part of the cities. Car rental agencies are plentiful, and you will need a car to visit most of the following points of interest.

Indoor Activities

Indoor activities in most cities are most appreciated by visitors during the chilling cold of winter. It is the enervating heat of summer that drives visitors indoors in Arizona.

Arizona Mineral Museum
17th Ave. & McDowell Rd., Phoenix, AZ 85009—602-255-3791

This museum features many industrial mineral displays, gemstones, and ores found in Arizona. It is open Monday–Friday, 8:00 A.M.–5:00 P.M. and Saturday, 1:00 P.M.–5:00 P.M.

Arizona Museum of Science and Technology
80 N. 2nd St., Phoenix, AZ 85004—602-256-9388

Hands-on exhibits of energy, life sciences, and the environment are displayed here. The museum is open Monday–Saturday, 9:00 A.M.–5:00 P.M. and Sunday, 1:00 P.M.–5:00 P.M.

Grace H. Flanrau Planetarium
University Blvd. & Cherry Ave., Tucson, AZ 85705—602-621-4515

Science halls and exhibits of astronomy and space exploration are included here. A 16-inch telescope is available for free night viewing, weather permitting. The planetarium is open Monday–Friday, 10:00 A.M.–4:00 P.M., Saturday–Sunday, 1:00 P.M.–4:00 P.M., and Tuesday–Saturday nights, 7:00 P.M.–9:00 P.M.

Kitt Peak National Observatory
950 N. Cherry Ave., Tucson, AZ 85719—602-620-5350

The observatory is in the Quinlan Mountains about 55 miles southwest of Tucson off State Route 86 near Sells, Arizona. The visitor center offers guided tours Saturday, Sunday, and holidays at 10:30 A.M. and 1:30 P.M. Exhibits on the universe and observatory equipment, which includes an 82-inch solar telescope and an 158-inch stellar telescope, are also available. The observatory is open daily from 10:00 A.M. to 4:00 P.M.

Mineral Museum
Geology Bldg., University of Arizona, Tucson, AZ 85705—602-621-4227

Excellent displays of gems, fossils, and minerals from around the world are exhibited here. The museum is open Monday–Friday, from 8:00 A.M. to noon.

Outdoor Activities

Outdoor acitivities are most popular in Arizona during the fall, winter, and spring, but many can be enjoyed in the summer. Desert drives are often pleasant in the

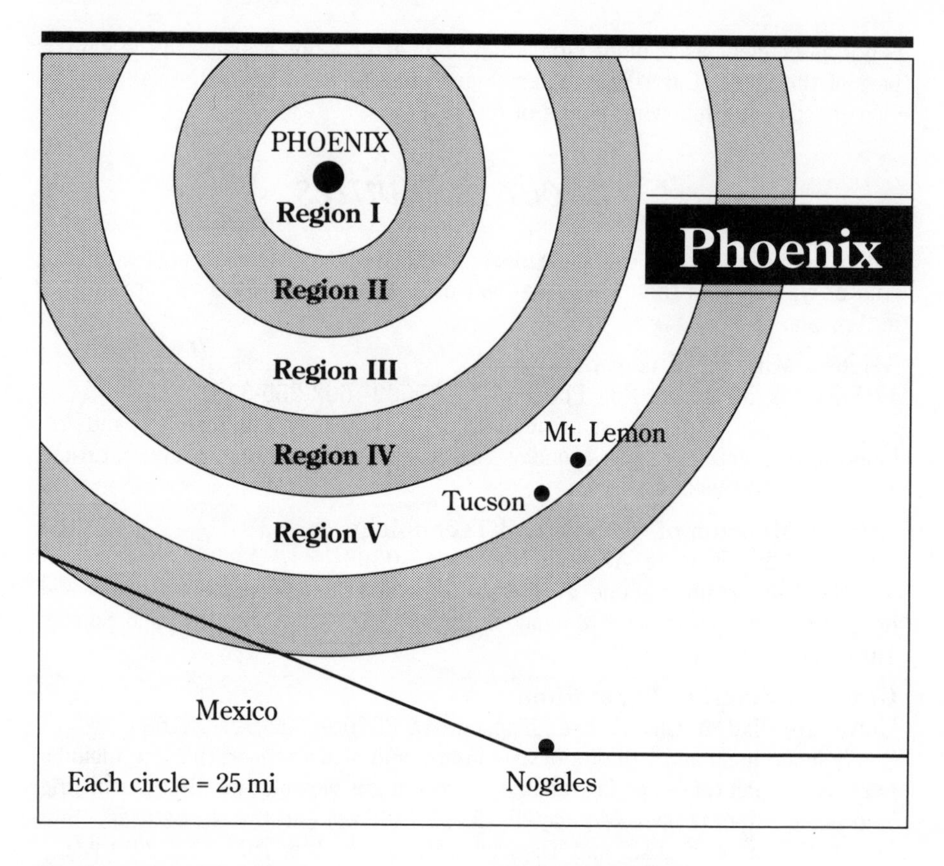

Indoor Activities

Region I
Arizona Mineral Museum
Arizona Museum of Science and Technology

Region V
Grace H. Flanrau Planetarium
Kitt Peak National Observatory
Mineral Museum

Outdoor Activities

Region I
Desert Botanical Garden
Echo Canyon Park
Encanto Park
Estrella Mountain Regional Park
Lake Pleasant Regional Park
North Mountain Park
South Mountain Park
Squaw Peak Park

Region II
Apache Trail

Region III
Boyce Thompson Southwestern Arboretum
Lost Dutchman State Park
Picacho Peak State Park
Tonto Natural Bridge

Region V
Catalina State Park
Sabino Canyon
Saguaro National Monument, Rincon Mountain Unit
Tohono Chul Park
Tucson Botanical Gardens
Tucson Mountain Park

cool of the morning and evening. Most parks are open until midnight so visitors can use them year-round.

Apache Trail

The road from Apache Junction to Globe on State Route 88 parallels the ancient route the Apache Indians used through the canyons of the Salt River. You climb past the Superstition Mountains, home of the famous Lost Dutchman Gold Mine. This drive offers changing vistas at each turn of the narrow and winding road. It is gravel for a 25-mile section between Apache Trail and Tortilla Flat. Among the many scenic views on the route is Fish Creek Canyon, which is noted for the massive, brightly colored walls that rise as much as 2,000 feet above the road. This 76-mile road is not a true nature outing, but visitors can get an understanding of the great variety of sites seen in the southwest deserts. Several of the following points of interest are located near it.

Bicycle Trails

Almost 50 miles of bike trails have been built in and around Phoenix. A detailed map and further information on the trails is available from the Phoenix Department of Parks and Recreation, 2333 North Central Avenue, Phoenix, AZ 85004; 602-262-7660.

Boyce Thompson Southwestern Arboretum

Superior, AZ 85273—602-689-2811

William Boyce Thompson, a local mining magnate, started this collection of cacti and other hot-weather plants during the 1920s. Today the 1,000-acre site is an outstanding botanical garden and research center with over 1,500 plant species from around the world. More than two miles of trails wind through the gardens, and some 70 species of animals and 175 of birds are often seen among the plants. While some plants are in bloom almost year-round, the best time to visit the arboretum is between October and May. Summertime temperatures can reach near 120°F. The arboretum is open daily and is on State Route 60, West of Superior.

Catalina State Park

P.O. Box 36986, Tucson, AZ 85740—602-628-5798

During the spring, Mexican goldcup poppies make a carpet of gold on the hills of this park. Visitors can explore the 5,500 acres of foothill and canyon land that varies in elevation between 2,600 and 3,000 feet on a variety of trails. One nature trail is an easy Sunday stroll, while other more difficult trails lead from the upper end of the park into the Santa Catalina Mountains of the million-acre Coronado National Forest. This range rises to 9,157 feet atop Mount Lemmon, and the rugged landscape is home to a wide assortment of wildlife and flora. Over 100 species of birds have been seen here, and the spring wildflower bloom is often spectacular.

Desert Botanical Garden

1201 N. Galvin Pkwy., Phoenix, AZ 85008—602-941-1225

Located in Papago Park, this 150-acre garden is devoted to arid land plants of the world. From late March to May the cactus blooms offer a colorful site. The garden is open daily from 7:00 A.M. to sunset.

Echo Canyon Park
East McDonald Dr. & Tatum Blvd., Phoenix, AZ 85018—602-262-7660

This is a small park tucked into the side of Camelback Mountain, with trails that lead through a wide variety of desert flora. It is open daily from 5:30 A.M. to midnight.

Encanto Park
North 15th Ave. & West Encanto Blvd., Phoenix, AZ 85007—602-262-7660

This large urban park has a lagoon that is a waterfowl refuge, unusual trees and shrubs, and nature trails. It is open daily from 5:30 A.M. to 12:30 A.M.

Estrella Mountain Regional Park
c/o Phoenix Department of Parks and Recreation, 2333 N. Central Ave., Phoenix, AZ 85004—602-932-3811

This park contains 18,600 acres of rugged desert terrain. A number of hiking trails lead visitors through native vegetation. The park is southwest of Phoenix off Interstate 10.

Lake Pleasant Regional Park
c/o Phoenix Department of Parks and Recreation, 2333 N. Central Ave., Phoenix, AZ 85004—602-566-0405

This large, 14,400-acre park has two lakes and many forms of desert vegetation, including the giant saguaro cactus. It is off Interstate 17 about 30 miles north of Phoenix.

Lost Dutchman State Park
Apache Junction, AZ 85220—602-982-4485

Off State Route 88, or The Apache Trail, and east of Phoenix, the park is named after the legendary gold mine. It is popular in all but the summer months for naturalists who want to hike in and around the Superstition Mountains. While the park is only 300 acres, with camping and picnicking, it is a convenient point of departure for hikers and horse riders who want to explore the 124,117-acre Superstition Mountains Wilderness Area.

For the less energetic, there are several short nature study trails within the park. For the more adventurous, there are several well-marked trails leading into the rugged and inhospitable wilderness. There is little water in the backcountry. Visitors wishing to explore there should consult with local rangers at the Mesa District Office of the Tonto National Forest about conditions and maps. Mesa District Office, Tonto National Forest, P.O. Box 5800, Mesa, AZ 85211—602-835-1161 or 602-835-5800

North Mountain Park
10600 N. 7th St., Phoenix, AZ 85020—602-943-5944

One of the smaller desert parks in the Phoenix area at 295 acres, this park has several nature trails that traverse a scenic mountainous area. It is open daily from 5:00 A.M. to 12:30 A.M. Rangers are on duty Tuesday–Friday, 1:30 P.M.–10:00 P.M. and Saturday–Monday, 5:00 A.M.–10:00 P.M.

Sabino Canyon

Sabino Canyon Rd. & Sunrise Dr., Tucson, AZ 85715—602-749-2861

East of Tucson in the Santa Catalina Mountains, this is a scenic canyon of geological interest. Desert plants dot the desolate canyon walls and lush growth covers the canyon floor along the Sabino Creek. There is a visitor center on top, and several trails lead into the canyon. No motor vehicles, other than the shuttle buses provided by the park, are allowed. The Bear Canyon Trail leads into a remote section of the canyon where a number of waterfalls are located. One of the most beautiful times to visit the canyon is at night during a full moon. The shuttle bus runs special trips at night the last three days before every full moon. The visitor center is open daily from 8:00 A.M. to 4:30 P.M.

Saguaro National Monument, Rincon Mountain Unit

P.O. Box 17210, Tucson, AZ 85731—602-629-6680

This site is east of Tucson off State Route 83. An eight-mile loop drive and a one-mile nature trail lead visitors through this monument to the giant saguaro cactus. These can reach over 50 feet in height, and live over 200 years. Numerous species of wildlife feed on, and live in and around these giants. Deer, coyotes, peccaries, owls, roadrunners, and woodpeckers are often seen in the monument. Clusters of white blossoms turn the tips of the saguaro's branches a light cream during May and early June. Bright red fruit follows these during late June.

For visitors who wish to enjoy the backcountry, more than 50 miles of hiking and horseback-riding trails wind through a 58,000-acre wilderness area that includes the fir-covered slopes of the Rincon Mountains. These reach over 8,700 feet in elevation.

There is another smaller unit of the Saguaro National Monument at Tucson Mountain Park. See entry in this section.

Picacho Peak State Park

P.O. Box 275, Picacho, AZ 85241—602-466-3183

This 3,400-acre park, off Interstate 10, is noted for its brilliant spring poppy blossom. It also offers visitors year-round opportunities to explore the desert habitat around the 3,374-foot peak. The Hunter Trail to the summit is a steep climb that takes four to five hours round trip. A shorter version, which stops at the upper saddle to the summit, only takes about two hours. Other trails lead to the low saddle and to Children's Cave.

A self-guided nature trail loops out of the campground for those who do not wish to stray too far into the desert. A brochure, available at the entrance station, includes maps of all the trails. Visitors are warned to keep a watch out for rattlesnakes and scorpions while in the park.

South Mountain Park

c/o Phoenix Department of Parks and Recreation, 2333 N. Central Ave., Phoenix, AZ 85004—602-276-2221

A large desert park within an easy drive of downtown Phoenix, South Mountain has more than 17,000 acres of peaks, canyons, exotic rock formations, and vegetation found only in Arizona. There is a road to the top of South Mountain. Dobbins Lookout, at 2,330 feet, offers an excellent view of Phoenix and the surrounding area. A number of hiking trails also lead to other viewpoints. The park is open daily from 5:30 A.M. to midnight.

Squaw Peak Park
2701 E. Squaw Peak Dr., Phoenix, AZ 85016—602-262-7660

A 1.2-mile summit trail is the feature of this park, which sits north of Phoenix. The trail offers an excellent view of the valley, and winds through many types of desert flora. Wildlife, both large and small, may be seen during early morning and evening hikes. The park is open daily from 5:30 A.M. to midnight.

Tohono Chul Park
7366 N. Paseo del Norte, Tucson, AZ 85704—602-742-6455

This is a small preserve where cacti and other succulents can be observed in a natural desert environment. Small desert animals are often seen, especially during the cooler hours. Indoor exhibits on desert life are featured, and geological specimens can be seen in the Demonstration Garden. The park is open daily from 7:00 A.M. to sundown. The exhibit building is open daily from 9:30 A.M. to 5:00 P.M.

Tonto Natural Bridge
Payson, AZ 85541—602-476-3440

The bridge is off State Route 87, 11 miles north of Payson. It is a natural bridge, purported to be one of the largest natural travertine structures in the world. It reaches 183 feet high, 400 feet long, and has an opening below that is 150 wide. A trail leads from the top of the bridge to the canyon below. Caves in the canyon walls contain traces of prehistoric habitation. The trail is steep and difficult.

Tucson Botanical Gardens
2150 N. Alvernon Way, Tucson, AZ 85712—602-326-9255

These are small gardens with arid, semi-arid, and tropical plants displayed as landscape for an adobe home. They are open weekdays, 9:00 A.M.–4:00 P.M. and weekends, 10:00 A.M.–4:00 P.M.

Tucson Mountain Park
Speedway Blvd. & Kinney Rd., Tucson, AZ 85746—602-742-6455

The 17,000 acres in this park include the Tucson Mountains and adjoining mesa lands. One of the largest areas in the Southwest of giant saguaro cactus with natural desert growth, the park has a visitor center and many hiking trails. It is open daily from 7:00 A.M. to 5:30 P.M.

Organizations That Lead Outings

Friends of Arizona Highways
2039 W. Lewis Ave., Phoenix, AZ 85009—602-258-6641

Sierra Club—Southwest Office
3201 N. 16th St., Phoenix, AZ 85016—602-277-8079

Tucson Audubon Society
300 E. University Blvd., Tucson, AZ 85705—602-629-0510

Wilderness Society
234 N. Central Ave., Phoenix, AZ 85004—602-256-7921

A Special Outing

When the British Broadcasting Corporation filmed a special documentary on the seven top zoos of the world, the Arizona–Sonora Desert Museum outside Tucson was included. It was second only to the San Diego Zoo in the United States.

Not a common zoo, this reserve presents an excellent overview of the unique arid region that is found in the southwest desert and extends into Mexico. It has over 500 species of animals and plants that are native to the Sonoran Desert of Arizona, Sonora, Mexico, and the Gulf of California. All are housed in native settings.

To reach the museum you take a scenic 14-mile drive west of downtown Tucson. You cross the 17,000-acre Tucson Mountain Park through forests of giant saguaro cactus and desert mountains.

Once you get there, you can study the flora and fauna of the region in a walk-in aviary where desert birds live in a natural setting. You may study earth science in a maze of artificial but very realistic wet and dry limestone caves, and look through large windows into the dark world of nocturnal desert life.

Above ground there are a variety of large desert mammals and reptiles living in natural settings, including a large prairie dog village. The zoo is open daily from 7:30 A.M. to 6:00 P.M. between Memorial Day and Labor Day. It is open from 8:30 A.M. to 5:00 P.M. during the rest of the year. For more information phone 602-883-1380. The zoo is located at 2021 North Kinney Road, Tucson, AZ 85746.

Nature Information

Arizona Game and Fish Commission
2222 W. Greenway Rd., Phoenix, AZ 85023—602-942-3000

Arizona Office of Tourism
1100 W. Washington St., Phoenix, AZ 85007—602-255-3618

Arizona State Parks
800 W. Washington St., Suite 415, Phoenix, AZ 85007—602-542-4174

Metropolitan Tucson Convention and Visitors Bureau
450 West Paseo Redondo, Suite 110, Tucson, AZ 85705—602-624-1817

Phoenix and Valley of the Sun Convention and Visitors Bureau
4455 E. Camelback Rd., Bldg. D, Phoenix, AZ 85018—602-254-6500

Phoenix Department of Parks and Recreation
2333 N. Central Ave., Phoenix, AZ 85004—602-262-7660

Tucson Department of Parks and Recreation
900 S. Randolph Way, Tucson, AZ 85716—602-791-4873

U.S. Forest Service
Southwestern Region, 517 Gold Ave., SW, Albuquerque, NM 87102—505-842-3292

Further Reading

Annerino, John. *Outdoors in Arizona: A Guide to Hiking and Backpacking.* Phoenix: Arizona Highways, 1987.

Ayer, Eleanor. *Arizona Traveler—Birds of Arizona: A Guide to Unique Varieties.* Washington, DC: R H Publications, 1988.

Chronic, Halka. *Roadside Geology of Arizona.* Missoula, MT: Mountain Press, 1983.

Heylmun, Edgar G. *Guide to the Santa Catalina Mountains of Arizona.* Tucson: Treasure Chest, 1979.

Lowe, Charles H. *Arizona's Natural Environment: Landscapes and Habitats.* Tucson: University of Arizona Press, 1972.

Merriam, Robert. *Arizona Minerals and How to Find Them.* Tucson: Treasure Chest, 1988.

Panczner, Sharon, and Bill Panczner. *Arizona Traveler—Gems and Minerals of Arizona: A Guide to Arizona's Native Gemstones.* Washington, DC: R H Publications, 1988.

Perry, John, and Jane G. Perry. *The Sierra Club Guide to the Natural Areas of New Mexico, Arizona, and Nevada.* San Francisco: Sierra Club Books, 1986.

Smith, Deborah. *Arizona Traveler—Arizona Cactus: A Guide to Unique Varieties.* Washington, DC: R H Publications, 1988.

Snyder, Ernest E. *Arizona Outdoor Guide.* Longview, AZ: Golden West Publishers, 1985.

FAR WEST

Seattle
San Francisco
Los Angeles
Honolulu

As with the Southwest, there is much diversity among the cities represented in the Far West, but there are also some striking similarities. All sit on or near the Pacific Ocean, and each is affected in its own way by the sea. Seattle lies somewhat inland, and is separated from the Pacific by the great Puget Sound. The Sound helps moderate the influence of the open sea, whose strong winds and cold Arctic fronts become weaker as they cross it. San Francisco adjoins the open ocean, but its residents and visitors are generally more aware of the San Francisco Bay. The high fogs that so strongly affect the unusual summer climate of the region are a peculiar product of the confluence of the open sea with the Sacramento Delta and San Francisco Bay. Los Angeles is distinguished by its Mediterranean climate that results from interaction of the desert and the sea. Honolulu, along with the rest of the Hawaiian Islands, cannot be separated from the dramatic influences of the mid-Pacific.

The cities of the West also share the seismic activity associated with tectonic plate movement. Seattle sits in an area with a number of small tectonic plates whose movements helped form the volcanoes of the Cascade Mountains. This movement continues, as demonstrated by moderate earthquakes in the Seattle region during the past several decades, and by the violent eruption of Mt. St. Helens some 100 miles south of the city in 1980. The activity along the plates of the Northwest will continue, with small to moderate quakes projected. There is a possibility of further volcanic activity in the Cascades over the next century.

San Francisco lies astride the fault line between two major tectonic plates. As recent history has demonstrated, the city is subject to large, even great, earthquakes. The 7.1 earthquake in October, 1989, was only the latest quake to shake the San Francisco region. As strong as it was, it was far surpassed by the 1906 temblor, which is estimated as 8.3 on the open-ended Richter scale, rated as a great quake.

All of this activity is a result of the movement of two massive tectonic plates that meet at the San Andreas Fault. To the east of the fault lies the North American Plate, and to the west, the Pacific Plate.

The Pacific Plate is moving northward causing large amounts of stress to build up along the fault. The Point Reyes Peninsula lies on the Pacific Plate west of the San Andreas Fault. During the past million years or so, it has moved from the southern California coast to its present location.

San Francisco is not the only city threatened by this movement. In the next 50 years, Los Angeles is probably in greater danger than San Francisco of experiencing a great quake, one above 8 on the Richter scale.

The San Andreas extends more than 750 miles off the northwestern coast of California to San Francisco, then through the mountain ranges just north of Los Angeles, where it turns inland. The combined stresses along the San Andreas and several other large faults in southern California make Los Angeles a prime target for disastrous temblors.

The Hawaiian Islands owe their very existence to seismic activity. As the Pacific Plate makes its journey northward it passes over a hot spot in the earth's mantle. Each island of the Hawaiian chain has been formed in turn over that hot spot. The big island of Hawaii now sits over that hot spot, and is the youngest island in the chain. It is also the most seismically active, with two volcanoes that produce numerous earthquakes as well as varying amounts of hot lava.

Visitors to the region soon accept this seismic activity as just another aspect of the natural environment of the Far West. They enjoy exploring it along with many other naturalist activities.

Among these, Seattle is noted as a starting point for explorations of the Puget Sound and San Juan Islands to the west, and of the majestic mountains of the Cascades and Olympics to the east and south. Most of these activities take visitors away from the city into rugged wilderness regions untamed by humans.

San Francisco, on the other hand, is more a central point than a starting point. Visitors can reach natural sites within minutes of downtown in every direction but the west. While they may not all be wilderness, since they are within easy reach of almost eight million people, they are refuges in the true meaning of the word.

Los Angeles has enveloped vast areas of desert and mountain land as it has grown to the second largest metropolitan area in the nation. It is difficult to define its relationship with nature. Some of the activities that are within a reasonable distance of downtown Los Angeles, which is even more hard to define since no one really knows where downtown is, are natural only in the sense that they are associated with nature. Other areas, most farther out in the desert and mountains, but some within the city limits of Los Angeles are still wilderness.

Honolulu is on an island that once had many interesting natural areas, but few have been preserved in a wilderness state. Tourism was unchecked for many years and overran just about everything natural. The other islands of Hawaii, however, have retained large areas of tropical wilderness.

Seattle

Seattle is known as the "Emerald City" because of its idyllic setting. It sits between the Puget Sound and Lake Washington, and the area's abundant rainfall and moderate climate keeps hills and meadows green year-round.

Water, both that of the Sound and that which falls from the heavens, contribute to the natural opportunities that can be enjoyed by visitors to Seattle. For those interested in sailing or diving the region is a veritable heaven. The 140-mile Puget Sound is one of the world's great embayments, and is filled with coves, bays, straits, islands, and sounds. These offer unlimited opportunities for sailors to explore, and the Sound offers some of the best cold-water diving in the world. A number of underwater parks are located throughout the region. There are also uncounted diving sites that are not designated as underwater parks, but offer excellent diving.

ABOVE: Mt. Rainier, which stands high above Western Washington, is an easy day-trip from Seattle. Here a hiker walks along the Mt. Rainier and Shadow Lake Trail. Photo by National Park Service

For those who are more interested in land-based exploration, Seattle offers great bird watching, thousands of miles of hiking trails within a 100-mile radius, and access to three of the most beautiful national parks in the country—Mount Rainier, North Cascades, and Olympic.

Add to this the tremendous positive attitude toward nature and the outdoors that is held by the general population, and you have an unmatched setting for visitors who are interested in naturalist activities. This positive attitude has led to the development of over 100 state parks, 26 wilderness areas, three national parks, six national forests, and three national recreation areas in Washington, many of which are within an easy drive from Seattle.

Climate and Weather Information

Seattle is infamous for its rainfall, yet its yearly total is only about 37 inches. What makes the amount seem much higher is the fact that over 80 percent of it falls between October and April, making those months quite wet. Summer has a large number of foggy and cloudy days, as well as some rain, that add to the impression of wetness.

While these two factors combine to make Seattle appear wetter than it is, the average year has only 47 clear, sunny days, as opposed to 221 cloudy or overcast ones. Measurable precipitation falls on 165 days.

While the area is definitely not Phoenix, where there are over 300 days of sunshine a year, Seattle does have moderate temperatures, averaging in the 70s during the summer and in the mid-40s during the winter. The city gets less than three inches of snow a year, and that generally melts off within a day.

Visitors should always be prepared for at least damp weather that is on the cool side. Rain gear is always a necessity, and waterproof shoes and boots make exploring the outdoors more comfortable.

Getting Around Seattle

It is not easy to drive in Seattle. The streets are supposed to run in a grid pattern, but they often veer off in unexpected directions. Street designations, such as N, S, E, SE, and SW, make things even more confusing.

Most of the outings are away from the center of town, however, and a rental car is almost a necessity. There are plenty of car rental agencies in the city, but it might be worthwhile to take public transit or a taxi to the outskirts of town before renting.

The transit system is good and covers downtown and the suburbs well. Parking in the downtown area is sparse, and taxis must either be obtained at a designated taxi stand or requested by phone.

The Washington State Ferry system offers inexpensive connections to the major islands in the Puget Sound.

Indoor Activities

These indoor activities are a welcome relief from the cold, wet weather that often comes to Seattle. Many visitors may wish to avoid it for at least short periods.

Hiram Chittenden Locks
3015 NW 54th St., Seattle, WA 98107—206-783-7001

These locks connect Puget Sound with Lake Washington and have fish ladders for migrating salmon and trout. There are six lighted windows below the water level that offer views of the migrating fish. The grounds are open daily from 7:00 A.M. to 9:00 P.M. The visitor center is open daily from 11:00 A.M. to 8:00 P.M.

Next door, the Carl S. English, Jr. Gardens have seven acres of luxurious plantings. Visitors can obtain a self-guiding brochure of the gardens at the interpretive center at the locks.

Mountaineers Museum
719 Pike St., Seattle, WA 98101—206-284-6310

This museum features displays of climbing equipment and its evolution. It is open Monday–Friday from 9:00 A.M. to 5:00 P.M.

Museum of Natural History
Thompson Hall of Science, University of Puget Sound, 315 N. Stadium Way, Tacoma, WA 98403—206-756-3100

This museum has collections and exhibits of the Northwest. It is open Tuesday–Saturday, 9:30 A.M.–5:00 P.M. and Sunday, noon–5:00 P.M.

National Oceanic and Atmospheric Administration
7600 Sand Point Way, NE, Seattle, WA 98115—206-527-6046

This is the western regional center for the NOAA, and self-guided tours of the facilities are available during normal working hours. The center occasionally holds open house and free science education days. Call for more details and hours of operation.

Rock Chalet
Snoqualmie, WA 98065—206-434-6141

The largest and most complete private rock collection in the United States, plus the largest petrified wood collection in the world are located in the museum above the Rock Chalet. It is off Interstate 90 near Snoqualmie Pass. The museum is open daily from 9:00 A.M. to 6:00 P.M.

Seattle Conservatory
15th St. East & East Galer, Seattle, WA 98112—206-684-4043

Five separate, but connected, greenhouses offer visitors views of different collections of flowers and plants. Located in Volunteer Park, the conservatory

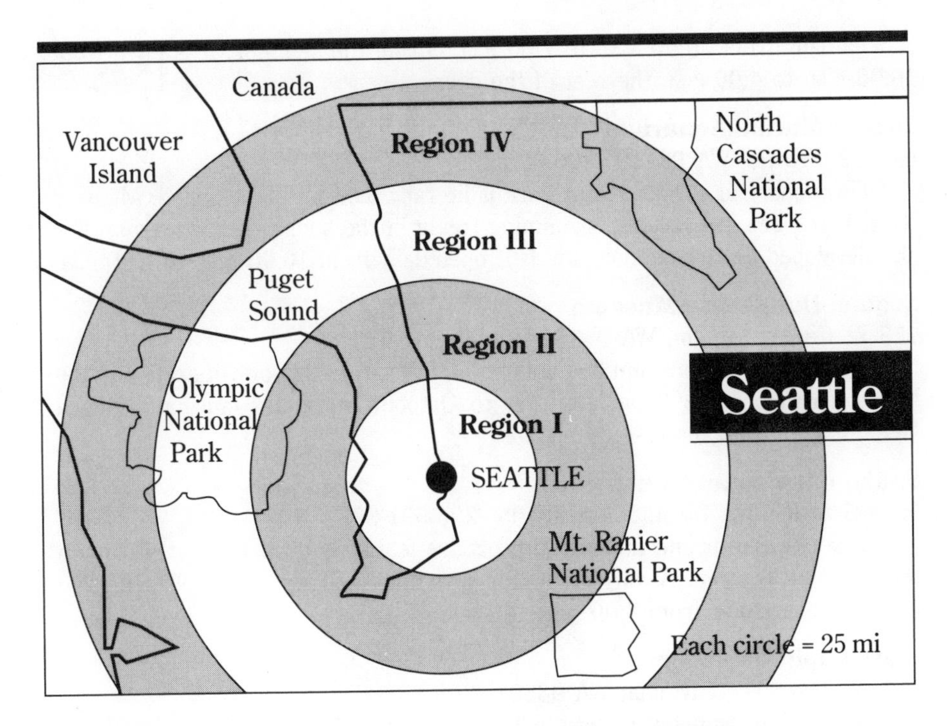

Indoor Activities

Region I

Hiram Chittenden Locks
Mountaineers Museum
National Oceanic and Atmospheric Administration
Seattle Conservatory
Seattle Marine Aquarium
Thomas Burke Memorial Museum
Washington State Fish Hatchery

Region II

Museum of Natural History
Rock Chalet
Sequim–Dungeness Museum
Snake Lake Nature Center
State Capitol Museum
Tokul Creek Fish Hatchery

Region IV

University of Washington Marine Laboratories
Whale Museum

Outdoor Activities

Region I

Christensen Greenbelt Park
McMicken Heights Park
Saltwater State Park
Sammamish River Trail
University of Washington Arboretum

Region II

Deception Falls Nature Trail
Dungeness National Wildlife Refuge
Flaming Geyser Recreation Area
Fort Nisqually Point Defiance Park
Marine Science Center
McLane Creek Nature Trail
Mima Mounds Preserve
Mount Si Preservation Area
Mount Walker Viewpoint
Nisqually National Wildlife Refuge
Nisqually Reach
Padilla Bay National Reserve
Skagit Habitat Management Area
Snoqualmie Falls
South Puget Sound Beaches
Tumwater Falls Park
Washington State Parks (see entries)

Region III

Baker Lake Recreation Area
Mt. Baker–Snoqualmie National Forest

Region IV

False Bay
Johns River Habitat Management Area
Sehome Hill Arboretum
Washington State Parks (see entries)

is open daily from 10:00 A.M. to 7:00 P.M. during the summer. It is open from 10:00 A.M. to 4:00 P.M. the rest of the year.

Seattle Marine Aquarium
Pier 59, Seattle, WA 98119—206-625-4357

This aquarium at Waterfront Park is literally built in Puget Sound. Many of the fish you see are actually swimming freely in the sound. The aquarium has also developed its own salmon run. It is open daily from 10:00 A.M. to 7:00 P.M.

Sequim-Dungeness Museum
175 W. Cedar, Sequim, WA 98382—206-683-6197

Exhibits at this museum include one on the natural history of area, and the Manis mastodon site. From May 1st to October 30th, the museum is open Wednesday–Sunday, noon–4:00 P.M.

Snake Lake Nature Center
1919 S. Tyler St., Tacoma, WA 98405—206-591-5939

A nature center and three loop trails are featured in this nature setting in the heart of Tacoma. A variety of free programs and tours are offered. The center is open daily from 8:00 A.M. to dusk.

State Capitol Museum
211 W. 21st Ave., Olympia, WA 98501—206-753-2580

Separate museums of history, natural history, and art, all focused on Washington are the featured attractions here. The museum is open Tuesday–Friday, 10:00 A.M.–4:00 P.M. and Saturday–Sunday noon–4:00 P.M.

Thomas Burke Memorial Washington State Museum
45th St. & 17th Ave., NE, Seattle, WA 98105—206-543-5590

The oldest university museum in the West, the Burke houses fine collections of anthropology, geology, and zoology. The observatory across the street was built in 1895, and uses a telescope manufactured in 1891. Both are still in use and offer public viewing. The museum is open Monday–Friday, 10:00 A.M.–5:30 P.M. and Saturday, 9:00 A.M.–4:30 P.M. It is located off 17th Avenue at north entrance to the University of Washington.

Tokul Creek Fish Hatchery
Snoqualmie, WA 98065—206-222-5464

Visitors can view steelhead and freshwater trout at the facility, as well as tour the hatchery. A picnic area is located nearby. The hatchery is off State Route 202 between Snoqualmie and Fall City.

University of Washington Marine Laboratories
Friday Harbor, WA 98250—206-378-2165

Free tours of the lab are given by graduate students in July and August on Wednesdays and Sundays at 2:00 P.M. and 4:00 P.M. The laboratories are located on San Juan Island.

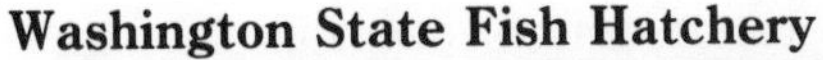

Washington State Fish Hatchery
125 W. Sunset Way, Issaquah, WA 98027—206-392-3180

The best time to visit this facility is between September 1st and October 31st when the salmon run is in progress. It is open daily from 8:00 A.M. to dusk.

Whale Museum
62 1st St., Friday Harbor, WA 98250—206-378-4710

Skeletons, sculptures, and information on new whale research are all available at this San Juan Island museum run by the Moclips Cetological Society. It is open daily from 10:00 A.M. to 5:30 P.M.

Outdoor Activities

Seattle is blessed with some of the most diversified outdoor activities of any city in this guide. From scuba diving in the Puget Sound to ice climbing on Mount Rainier there is some outdoor activity for everyone.

Baker Lake Recreation Area
Baker River Ranger District, U.S. Forest Service, Concrete, WA 98237—206-853-2851

Off State Route 20 north of Concrete, Washington, this nine-mile human-made lake sits at the base of Mount Baker. It offers visitors the essence of the Pacific Northwest. Dense rainforests, snowcapped mountains, and wooded valleys with highland lakes are all found nearby. The Shadow of the Sentinels Nature Trail near the main entrance to the recreation area gives visitors a good introduction to the region's plant life, and at the north end of the lake, agates and jaspers can be found at the mouth of Swift Creek.

Bicycle Trails

In all, Seattle has nearly 100 miles of official bike trails, the most popular one being the Burke Gilman Trail. It runs from Gasworks Park on Lake Union to Kenmore Logboom Park at the northernmost point of Lake Washington. The Burke Gilman Trail is part of a series of biking and hiking trails in the state of Washington that follow old railroad right-of-ways, and now total over 150 miles. There are plans for these trails to reach nearly across the state.

For more information on bike trails contact the Seattle/King County Visitor and Convention Bureau, 1815 7th Avenue, Seattle, WA 98101; 206-447-4240 or the Seattle Parks and Recreation Department, 520 Green Lake Way, North, Seattle, WA 98103; 206-684-4075.

Blake Island State Park
P.O. Box 287, Manchester, WA 98353—206-447-1313

Over four miles of beaches are a main feature of this island park in Puget Sound. It is accessible only by boat or ferry. A hiking trail circles the island which was once a private estate. It is now densely wooded with mostly native plants. There is also a three-quarter-mile nature trail. Other trails total over eight miles. Scuba divers enjoy the underwater park area. A developed area includes Tillicum Village, where Indian dances and salmon barbecues are held during the peak tourist season.

Camano Island State Park
2269 S. Park Rd., Stanwood, WA 98282—206-675-2417

This island park, which is accessible by car, is located west of Stanwood on State Route 532 in the Saratoga Strait between Whidbey Island and the mainland. It is popular for its bird watching and clamming. Several trails lead through the forest that covers about 85 percent of the island. Bald eagles winter here, as well as many waterfowl and shorebirds.

Christensen Greenbelt Park
c/o King County Parks Department, King County Courthouse, 516 3rd St., Seattle, WA 98104—206-344-4232

The 1.6-mile trail along the bank of the Green River offers visitors a chance to hike along the shore. Bird watching is also a popular activity here. The park is in Tukwila, Washington.

Deception Falls Nature Trail
Skyhomish Ranger District, U.S. Forest Service, Skyhomish, WA 98288—206-677-2414

At Deception Creek a half-mile nature trail skirts the Tye River and offers views of two waterfalls. A self-guided trail brochure explains 15 points that are marked along the trail.

Deception Pass State Park
5175 N.S.H. 20, Oak Harbor, WA 98277—206-675-2417

North of Oak Harbor on State Route 20, this popular state park has great variety. There are both sandy and stony beaches, tide pools, rocky headlands, coves, bays, cliffs, wetlands, coastal brush, and forests. Local bird-watchers recommend the area because of its diverse habitat.

Orca, or killer whales, and other whales are often spotted from the headlands. Bald eagles, osprey, and other raptors are seen regularly. The park has many trails in various sections, and these lead to a wide variety of habitats.

Dungeness National Wildlife Refuge
c/o Nisqually National Wildlife Refuge, 100 Brown Farm Rd., Bldg. A, Olympia, WA 98506—206-753-9467

The refuge northwest of Sequim is on a natural sand spit that projects over five miles into the Strait of San Juan de Fuca. It is a favorite spot for bird watching because it has such a large number of species that live in, or migrate through, the area.

Sequim is in the rain shadow of the Olympic Mountains, and receives less than 18 inches of rain a year. This makes the surrounding area popular when others are wet and chilly.

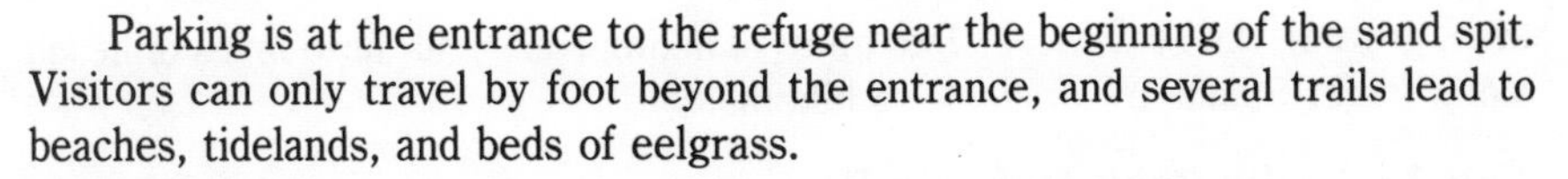

Parking is at the entrance to the refuge near the beginning of the sand spit. Visitors can only travel by foot beyond the entrance, and several trails lead to beaches, tidelands, and beds of eelgrass.

False Bay

During spring and summer a walk along the northwestern shore of this bay on San Juan island leads visitors near nesting bald eagles. Even at other times of the year the bay is an almost foolproof spot for eagle watching. The whole bay is a University of Washington biology preserve and the site of numerous student research projects.

Federation Forest State Park

Star Route, Enumclaw, WA 98022—206-663-2207

Visitors to Mount Rainier National Park often stop here. The 609-acre site of mostly virgin timber was acquired through the efforts of the Washington State Federation of Women's Clubs. Funds willed to the Federation were used to build the Catherine Montgomery Interpretive Center. Exhibits here describe the flora and fauna of the principal life zones of the state, and large windows offer visitors opportunities to look out onto small gardens with living specimens mentioned in the exhibits. There are two outdoor nature trails. The park is southeast of Enumclaw on State Route 410.

Flaming Geyser Recreation Area/Nolte State Park
and
Green River Gorge Conservation Area

23700 SE Flaming Geyser Rd., Auburn, WA 98002—206-931-3930

North of Enumclaw on State Route 169, the Green River Gorge is a scenic area heavily used by a number of groups. White water rafting is popular during the winter and early spring. Hunters and fishermen also use the area during season.

Heavy forests with a variety of wildflowers cover about 60 percent of the area. There is also a small swamp and several lakes. Trails include the one-mile River Trail that follows the river upstream into the Gorge, and the one-and-one-half-mile Hill Trail that leads to the top of a wooded hill above the day-use area. There is a small lake and a self-guided nature trail at Nolte State Park.

Fort Casey State Park

1280 S. Fort Casey Rd., Coupeville, WA 98239—206-678-4519

An interpretive center on the beach and an underwater reserve for scuba divers are the two main features of this small park next to the Keystone Ferry landing. The park also provides access to most of the beachfront along the west shore of Whidbey Island all the way to Fort Ebey State Park.

Fort Ebey State Park

395 N. Fort Ebey Rd., Coupeville, WA 98239—206-678-4636

This is a new park located in the driest spot on Whidbey Island. Prickly pear cacti, more often connected with the southwest desert than with the Puget Sound area, grow on the bluffs here. Two miles of hiking trails and three miles of beach along Puget Sound are available to hikers.

Fort Nisqually Point Defiance Park
North 54th & Pearl Sts., Tacoma, WA 98407—206-759-1246

This was originally a 640-acre military reserve. There are over 700 acres in the park, and maintained trails lead through the rhododendron gardens and other portions of the park. A five-mile drive leads through 500 acres of virgin timber, and has a number of scenic viewpoints overlooking Puget Sound.

Hiking in Seattle

Hiking is easy in Seattle, and there are many trails in and around the city. These include the following.

Myrtle Edwards Park has a two-mile path along Elliot Bay. Foster Island Nature Walk is on a paved trail over marshlands beside Lake Washington, and Seward Park is a small peninsula on the south end of Lake Washington. Schmitz Park is a wooded area in west Seattle, and Discovery Park is a former military base with over two miles of beach. For maps and trail information, contact the Seattle Parks and Recreation Department, 520 Green Lake Way, North, Seattle, WA 98103; 206-684-4075.

Hiking in the wilderness areas outside Seattle is some of the best in the nation. Sections of the Cascade Crest and Pacific Crest Trails and park trails in Olympic, North Cascades, and Mt. Rainier National Parks are all within easy reach of Seattle. For complete trail information, contact the U.S. Forest Service–National Park Service Outdoor Recreation Office, 1018 1st Avenue, Seattle, WA 98104; 206-442-5400. Also see the Special Outing section for Seattle.

Johns River Habitat Management Area
905 E. Heron, Aberdeen, WA 98520—206-533-9335

The habitat is much the same as the Olympic National Park and Forest, and is home to Olympic elk, river otters, coyotes, raccoons, and mule deer. A number of trails lead along the dikes and river. The area is about 12 miles southwest of Aberdeen on State Route 105.

Lake Cushman State Park
P.O. Box 128, Hoodsport, WA 98548—206-877-5491

This park is lightly used by a few hunters and fisherman. Hiking is easy along dikes on both sides of the lake, which was formed by a dam on the Skykomish River near the Olympic National Park. Tidelands, floodplains, and upland woodlands comprise most of the park, and a wide variety of birds can be spotted there. Upland game species such as blue and ruffed grouse, band-tailed pigeon, and Chinese pheasant can be seen, along with migrating waterfowl and raptors such as marsh, red-tailed, and Cooper's hawks. Roosevelt elk, black bears, and mule deer also reside in the park, as do smaller mammals such as muskrats, mink, coyotes, and river otters.

Marine Science Center

17771 Fjord Dr., NE, Poulsbo, WA 98370—206-779-5549

Educational exhibits and live marine specimens are featured here. The center is open Tuesday–Saturday, 9:00 A.M.–5:00 P.M. during July and August. It is open Monday–Friday, 8:00 A.M.–4:00 P.M. during the rest of the year.

Matia Island Marine State Park

Star Route, P.O. Box 22, Eastsound, WA 98245—206-376-2326

This island park is part of a larger national wildlife refuge and wilderness system. It is accessible only by boat. Public use is restricted to the northwest end of island, but trails lead up steep banks to forested areas. Matia Island Marine State Park is northeast of Orcas Island.

McLane Creek Nature Trail

c/o Thurston County Parks and Recreation Department, 529 W. 4th Ave., Olympia, WA 98501—206-786-5595

Visitors can learn about the local flora and fauna on this self-guiding mile-long trail. It is about five miles west of Olympia.

McMicken Heights Park

c/o King County Parks, King County Courthouse, 516 3rd St., Seattle, WA 98104—206-344-4232

Several trails lead through the thick forest of trees and ferns in this 11-acre natural area in Tukwila, Washington.

Mima Mounds Preserve

128 Ave., SW & Waddell Creek Rd., Olympia, WA 98501—206-753-2449

The 445-acres of strange earth mounds are a designated National Natural Landmark. Their origins have been studied for over a hundred years. No satisfactory explanation has been presented for their origin, but interpretive displays help visitors explore the possibilities. The Mima Mounds Preserve is southwest of Olympia off Interstate 5.

Moran State Park

Star Route, P.O. Box 22, Eastsound, WA 98245—206-376-2326

The 2,400-foot Mount Constitution with its 50-foot rock tower, and an abundance of game and bird life are features of this popular park on Orcas Island. Miles of trails lead through stands of virgin forest.

Mount Baker–Snoqualmie National Forest

c/o U.S. Forest Service–National Park Service Outdoor Recreation Information Office, 1018 1st Ave., Seattle, WA 98104—206-442-5400

There are over two million acres within the boundaries of this national forest that includes some of the wildest country in the northwest. It runs from the Canadian border to Mount Rainier along the west slope of the Cascade Mountains. Half of the federally owned 1.7 million acres is covered by virgin stands of timber.

Most of the area is within a reasonable distance of Seattle. Glacier Peak Wilderness is almost 500,000 acres with over 30 peaks. Four of these reach over 9,000 feet in elevation. Glaciers, hundreds of snow-fed mountain streams, and the Pacific Crest Trail are located in this wilderness area. Skagit Wild and Scenic River was designated in 1978, but still has no general management plan. The Alpine Lakes Wilderness has over 300,000 acres with more than 700 lakes, if those smaller than an acre are included.

Mount Si Preservation Area

North Bend Ranger District, North Bend, WA 98045—206-888-1421

This area is great for hiking with a number of good trails. These include Mount Si, Humpback Creek, Snow Lake, Rattlesnake Lake Ridge, Twin Falls, McClellon Butte, and Franklin Falls Trails. They all lead visitors through scenic areas of alpine timber land, jagged peaks, and mountain meadows.

Mount Walker Viewpoint

Quilcene Parks Department, Quilcene, WA 98376—206-765-3368

A hiking trail leads to top of 2,804-foot Mount Walker where visitors can gain a panoramic view of Puget Sound and the Olympic and Cascade Mountains. The viewpoint is about ten miles south of Quilcene.

Nisqually National Wildlife Refuge

100 Brown Farm Rd., Olympia, WA 98506—206-753-9467

This 3,780-acre refuge has estuaries, tidal flats, freshwater marshes, dikes, and grasslands that provide a wintering area for migratory waterfowl. Many other birds, including raptors, shorebirds, and songbirds consider this refuge home. It is also one of the few untouched river deltas in the country.

You can only gain access to the refuge on foot. Visitors can hike a trail to Twin Barns, follow the Nisqually River interpretive pathway, or explore the five-mile trail along the dike. The refuge is east of Olympia off Interstate 5.

Nisqually Reach

Audubon Nature Center, 4949 D'Milluhr Rd., NE, Olympia, WA 98506—206-459-0387

Evidence of glacial activity, seashore life, and a variety of birds, fish, and mammals all can be observed in this estuary area. More life is produced per acre in estuaries such as this than in any other natural environment. The Audubon Nature Center has interpretive displays that explain this to visitors.

Padilla Bay National Estuarine Research Reserve and Interpretive Center

1043 Bay View–Edison Rd., Bay View, WA 98257—206-428-1558

This is an 11,500-acre preserve that protects habitats such as open marine waters, tide flats, marshes, beaches, wooded uplands, and open fields. More than 250 bird species have been reported here, and an interpretive center explains the various habitats. There is also a marked nature trail.

Saltwater State Park
25205 8th Place, South, Des Moines, WA 98031—206-764-4128

This park is located on Puget Sound just below Kent Smith Canyon. It is south of Tukwila off State Route 509. Swimming, fishing, and scuba diving in the underwater park are favorite activities of residents. There are extensive trails through the park and canyon.

Sammamish River Trail
c/o King County Parks Department, King County Courthouse, 516 3rd St., Seattle, WA 98104—206-344-4232

This 12-mile trail is paved and offers many scenic stops. It is also connected to the Burke Gilman Trail for bicyclists. The trail is east of Kirkland.

Scuba Diving in Puget Sound

Puget Sound provides some of the best scuba diving in the nation. Underwater parks at Keystone, Edmunds, Deception Pass, Fort Worden, Kopachuck, Old Fort Townsend, Tolmie, and other areas give visitors opportunities to explore this activity in safe confines. Scuba gear can be rented in many shops in the area, but you must have verification of scuba certification with you to rent equipment. For more complete information on the underwater parks, contact the Washington State Parks and Recreation Commission, 7150 Clearwater Lane, Olympia, WA 98504; 206-753-5755.

Sehome Hill Arboretum
Western Washington University Campus, Bellingham, WA 98225—206-676-3000

This park overlooks Bellingham Bay, the San Juan Islands, and Mount Baker. Marked trails identify native plants. The arboretum is open daily during daylight hours.

Skagit Habitat Management Area
c/o Washington Department of Game, 600 N. Capitol Way, Olympia, WA 98504—206-753-5700

Near Conway off Interstate 5, this wildlife management area has about five miles of frontage on Skagit Bay. It includes extensive tide flats, sloughs, cattail salt marshes, wooded stream banks, lowland brush, and upland forest. It is reputed to be the most important waterfowl area in western Washington.

There are two large and several smaller tracts in the area. Visitors should check in at the headquarters on Mann Road for a map and advice on where to see the most waterfowl. There is a two-mile trail along the top of a dike near the headquarters. There are a number of other access points where thousands of waterfowl can be seen in the fall and spring, although many winter over, and some even use the area as nesting grounds during late spring and early summer.

Snoqualmie Falls
North Bend Ranger District, North Bend, WA 98045—206-888-1421

A spectacular rock gorge here has been carved out by the powerful Snoqualmie River. This and the 268-foot high Snoqualmie Falls are the principal attractions here.

South Puget Sound Beaches

There are over 23 state-owned beaches in the South Puget Sound area, but 21 are accessible only by boat. The Robert F. Kennedy Education and Recreation Area, two miles west of Longbranch, and Maple Hollow Beach, about five miles north of Longbranch, are accessible by car and offer visitors a chance to explore the area. The Washington State Department of Natural Resources has a publication titled *Your Public Beaches, South Puget Sound* that provides complete information on the area. Contact the DNR, Division of Marine Land Management, Olympia, WA 98504; 206-753-5324.

South Whidbey State Park

4128 Smuggler's Cove Rd., Freeland, WA 98249—206-321-4559

This park on Admiralty Inlet is comprised of 87 acres of virgin forest. The park features a good trail to a 4,500-foot beach.

Spencer Spit State Park

Star Route, P.O. Box 22, Eastsound, WA, 98245—206-376-2326

This heavily used 130-acre park has a large sand spit, shallow lagoons, sandy slopes, and forested hillsides. The island is readily accessible by ferry from Seattle. It is on the east side of Lopez Island.

Sucia Island Marine State Park

Route 2, P.O. Box 3600, Lopez, WA 98261—206-376-2326

This park, north of Orcas Island, consists of a group of islands arranged in a horseshoe with many bays and coves. These are some of the few boat-accessed islands in Puget Sound that offer hiking. With rocky beaches, eroded cliffs, and upland forests, the park offers visitors varied hiking opportunities and many seabirds.

Tumwater Falls Park

c/o Thurston County Parks and Recreation Department, 529 W. 4th Ave., Olympia, WA 98501—206-786-5595

Located next to the Olympia Brewing Company in Tumwater, Washington, this park has self-guiding trails that lead through rhododendron and azalea. Fantastic blooms are on display during the spring.

Turn Island Marine State Park

Star Route, P.O. Box 22, Eastsound, WA 98245—206-376-2326

This island park is part of a national wildlife refuge. It is close enough to San Juan Island that a canoe or rowboat crossing is feasible in good weather. It has beaches, tide flats, and woods.

University of Washington Arboretum

Lake Washington Blvd., Seattle, WA 98122—206-543-8000

The 250 acres in this arboretum have plantings that bloom year-round. A four-acre Japanese Tea Garden is the largest of its kind outside of Japan. The arboretum is open daily from 10:00 A.M. to 6:00 P.M.

Wallace Falls State Park
P.O. Box 106, Gold Bar, WA 98251—206-793-0420

This 518-acre park is located in the foothills of the Cascades, about 30 miles east of Everett near Gold Bar, Washington. It is a favorite of hikers. It features a two and one-half-mile trail to the falls that winds through a mixed, second-growth forest along the Wallace River. Viewpoints along the trail and at the top look out over the Skykomish River valley and the 250-foot Wallace Falls. The Old Railroad Grade Trail also goes to the top of the falls, and is often used as a return trail. It is about one mile longer, but has a more gentle grade than the Falls Trail.

Organizations That Lead Outings

Anchor Excursions
2500 Westlake Ave., North, Seattle, WA 98109—206-282-8368

Greenpeace
4649 Sunnyside North, Seattle, WA 98103—206-632-4326

Lake Union Air
1100 Westlake Ave., North, Seattle, WA 98109—206-284-0300

Northwest Rivers Council
P.O. Box 88 Main Office Station, Seattle, WA 98111—206-932-6587

Pacific Northwest Float Trips, P.O. Box 736, Sedro Woolley, WA 98284—206-855-0535

Rainier Mountaineering
535 Dock St., Suite 209, Tacoma, WA 98402—206-627-6242

Recreation Expeditions, Inc.
1525 11th Ave., Seattle, WA 98122—206-323-8333

San Juan Kayak Expeditions
3090-B Roche Harbor Rd., Friday Harbor, WA 98250—206-378-4436

Seattle Audubon Society
Joshua Green Bldg., Room 619, 1425 4th Ave., Seattle, WA 98101—206-622-6695

Sierra Club Northwest
1516 Melrose Ave., Seattle, WA 98122—206-625-0632

Urban Wildlife Coalition
P.O. Box 65027, Seattle, WA 98121—206-622-5260

A Special Outing

While visitors to New York, Minneapolis, and San Francisco have excellent access to nature outings, only visitors to Seattle have quick access to three major national parks. The North Cascades National Park is northeast of the city, Mount Rainier National Park is southeast, the Olympic National Park is southwest of Seattle. Each of these parks has its own distinctive characteristics, and each deserves extended visits to experience the wide variety of nature outings available.

The North Cascade National Park lies near the Canadian border and features unmatched alpine scenery with lakes, canyons, glaciers, and hundreds of jagged peaks. Established in 1968, this is one of the younger national parks of the West, and still relatively undisturbed by visitors and developers.

Deep, glaciated canyons cut through the Cascade Range beneath high peaks where over 300 glaciers can be seen. Waterfalls and icefalls drop out of hanging valleys, and deep blue lakes nestle in glacial cirques. Snow-fed streams cut through most of the wilderness.

Lakes large and small are found throughout the 1,053 square miles of back-country in the park, and trails lead to hidden valleys. For sheer alpine beauty the North Cascades National Park is matched only by the Rocky Mountain National Park near Denver, and the North Cascades Park is much less crowded.

South and east of Seattle, Mount Rainier rises 14,410 feet above sea level, and dominates the skyline for over 100 miles in every direction. The peak is an ice-clad, dormant volcano that has more glaciers than any other single mountain in the United States south of Alaska.

More than 300 miles of trails circle the mountain, where visitors can see glorious carpets of wildflowers late into the summer, as well as abundant wildlife. Mountain goats, deer, elk, and bears all can be spotted on even the more heavily traveled trails, especially those in the flower belt, which is above the 5,000-foot tree line.

Mount Rainier is one of the most popular mountain-climbing spots in the Northwest, but its unpredictable weather makes it dangerous. The National Park Services requires that all climbers on the upper reaches of the peak use a guide service registered with the park.

When it opened in 1899, Mount Rainier was the fourth national park designated by Congress. The park has continued to be a very popular destination for visitors to the region.

South and west of Seattle is the third major national park within driving distance. The Olympic National Park is a 1,400-square mile jumble of glacier-studded mountains, coniferous rain forests, and lakes.

The peaks are much lower than those in the North Cascades and Mount Rainier National Park. Only a few rise above 7,000 feet, and the highest is Mount Olympus at 7,965 feet. Nevertheless, these peaks are imposing since they literally rise from the sea.

Olympus National Park includes a 60-mile stretch of Pacific Ocean coastline with rocky headlands and sandy beaches. Its rain forests, where the Douglas fir and Sika spruce stand over 300 feet tall, have a yearly rainfall of 140 inches. The park also features more than 600 miles of trails that provide hikes which can last from several minutes to several weeks.

All of these parks have numerous visitor centers where you can find information on the region, and discover where you would like to explore. While these parks can be reached and examined in a cursory manner during a long-day outing, they are best enjoyed during a visit of at least several days. All offer campgrounds and lodging within the parks. For more information contact: Superintendent, North Cascades National Park, Sedro Woolley, WA 98284; 206-855-1331; Superintendent, Mount Rainier National Park, Star Route, Ashford, WA 98304; 206-569-2211, or Superintendent, Olympic National Park, 600 East Park Avenue, Port Angeles, WA 98362; 206-452-4501.

Nature Information

Anacortes Chamber of Commerce
1319 Commercial Ave. Anacortes, WA 98221—206-293-3832
Contact them for information on the San Juan Islands and northern Puget Sound.

Olympia Peninsula Tourism Council
P.O. Box 303 Port Angeles, WA 98362—206-479-3594

Seattle/King County Visitor and Convention Bureau
1815 7th Ave., Seattle, WA 98101—206-447-4240

Seattle Parks Department
Recreation Information Office, 520 Green Lake Way, North, Seattle, WA 98103—206-684-4075

Thurston County Parks and Recreation Department
529 W. 4th Ave., Olympia, WA 98501—206-786-5595

Travel Development Department
Department of Commerce and Economic Development, General Administration Bldg., Olympia, WA 98504—206-586-2088

U.S. Fish and Wildlife Service
Pacific Region 500 NE Multnomah St. Portland, OR 97232—503-231-6121

U.S. Forest Service
Northwest Regional Office, 319 SW Pine St., Portland OR 97208—503-221-2971 or 503-221-3644 for TRIS information

The USFS offers TRIS (Trail Information System), which covers the entire Puget Sound region, and offers specific information such as distances, difficulty, special conditions, restrictions, and prevailing weather for over 11,000 trails. Terminals can be found in public libraries and forest service offices.

U.S. National Park Service
Pacific Northwest Regional Office, 1424 4th Ave., Seattle, WA 98101—206-442-5565

U.S. Forest Service–National Park Service Outdoor Recreation Information Office
1018 1st Ave., Seattle, WA 98104—206-442-0170
Publications available from this office include *Wilderness Solitude Catalog, Pacific Crest National Scenic Trail,* and *The Campground Directory.*

Washington Department of Natural Resources
Public Lands Bldg., Olympia, WA 98504—206-753-5327

Publications available from this agency include *Washington State Outdoor Recreation Guide* and *Wildlife Recreation Areas,* plus some field guides to birds and mammals of the various regions.

Washington Department of Game
600 N. Capitol Way, Olympia, WA 98504—206-753-5700

Washington State Ferry System
State Ferry Terminal, Seattle, WA 98104—206-464-6400. For recorded schedules, call 800-542-0810 in Washington, 800-252-4550 outside Washington

Washington State Parks and Recreation Commission
7150 Cleanwater Ln., Olympia, WA 98504—206-753-5755

Further Reading

Furrer, Werner. *Water Trails of Washington.* Lynwood, WA: Signpost Publications, 1979.

Kirk, Ruth. *Exploring the Olympic Peninsula.* Seattle: University of Washington Press, 1975.

Larrison, Earl J. *Mammals of the Northwest.* Seattle: Seattle Audubon Society, 1976.

———. *Washington Wildflowers.* Seattle: Seattle Audubon Society, 1978.

Manning, Harvey, Bob Spring, and Ira Spring. *One Hundred Hikes in the South Cascades and Olympics.* Seattle: Mountaineers Books, 1985.

———. *Footsore 1, 2, 3, and 4.* 2nd. Ed. Seattle: Mountaineers Books, 1988.

Mueller, Marge. *San Juans Afoot and Afloat.* 2nd Ed. Seattle: Mountaineers Books, 1988.

———. *North Puget Sound, Afoot and Afloat.* Seattle: Mountaineers Books, 1988.

———. *South Puget Sound, Afoot and Afloat.* Seattle: Mountaineers Books, 1982.

Pyle, Robert M. *Watching Washington Butterflies.* Seattle: Seattle Audubon Society, 1974.

Schwartz, Susan, Bob Spring, and Ira Spring. *Wildlife Areas of Washington.* Seattle: Superior Publishing, 1981.

Spring, Ira. *One Hundred Hikes in the North Cascades.* Seattle: Mountaineers Books, 1988.

Sterling, E.M. *Trips and Trails 1 and 2.* Seattle: Mountaineers Books, 1983.

———. *Western Trips & Trails.* 2nd Ed. Boulder, CO: Pruett Publishing Co., 1987.

San Francisco

Called "Baghdad by the Bay" by its favorite columnist, San Francisco is more noted for its cosmopolitan nightlife than its natural wildlife. Yet, for all its reputation for dining and dancing, the city rates as one of the top cities in the country for the quality, quantity, and accessibility of its naturalist outings.

Bounded on the west by the Pacific Ocean, on the east by the San Francisco Bay, which is really an estuary, and to the north and south by the rolling to rugged hills of the Coast Range, nature is seldom far away from the city. Even national and state wilderness areas are less than an hour's drive from downtown.

Enjoyment of these attractions comes only with some understanding of how San Francisco and the central coast of California differ from other sections of the country. Many visitors have left after a summertime visit to San Francisco with nothing but sad memories of days spent in cold, foggy weather with nothing but shorts and light shirts to clothe themselves, when long pants and heavy

ABOVE: Visitors to the Marin Headlands can enjoy bird-watching, hiking, and a great view of the Golden Gate Bridge and downtown San Francisco. Photo by author.

jackets would have been more appropriate. Other visitors have left the region in an unsettled state, not because they did not enjoy the natural outings available, but because they could not put another natural phenomenon out of their mind—earthquakes.

The quality of naturalist's activities in and near the city is best indicated by the designation of the Central California Biosphere by UNESCO in 1989.

With a million acres of land and 30 square miles of ocean that stretch from Point Reyes in the north to the peninsula south of San Francisco, and west to the Farallon Islands, this reserve is one of almost 300 biospheres designated by the United Nations Educational, Scientific, and Cultural Organization. It is unusual among other reserves designated by UNESCO in that it is on the fringe of a major metropolitan area of close to six million people. While the reserve has no special authority, and the areas encompassed by it are not subject to any legal restrictions, it does indicate the effort that the region has undertaken to preserve its natural areas, and the progress it has made in balancing growth and preservation.

Climate and Weather Information

California offers many mysteries to visitors, and San Francisco's climate is just one more. This may be characterized in the oft-quoted statement attributed to Mark Twain that "the coldest winter I ever spent was a summer in San Francisco."

While it is debatable that Twain ever really made this statement, it is true that shorts and T-shirts are more appropriate during late September and October in San Francisco than they are during July and August. During those months, high fog frequently creeps in from the west and shrouds the higher peaks around the city to help keep daytime highs in the 50s for weeks on end.

As fall approaches the fog stays at sea, and highs reach into the 70s and low 80s for several months, until the winter storms reach down from the Gulf of Alaska. These bring a chill to the region, but temperatures seldom drop below freezing. Rain does come from these storms. While San Francisco itself only receives about 25 inches of rain from November to April, other cities in the region receive annual rainfall that varies from a high of about 60 inches to a low of about 15 inches a year. These variations are all within an hour's drive of downtown San Francisco.

With these variations, and with the lack of distinctive seasons, at least as far as visitors from east of the Rockies are concerned, San Francisco's climate offers many unusual sights. Green hills during the dead of winter, brown hills during the peak of summer, and roses and camellias in December.

Getting Around San Francisco

Downtown San Francisco is small in comparison to most major metropolitan areas. With a population of less than 800,000, and a very small land mass, nothing is very far from the center. Streets are often confusing as they follow the bay, and parking is pretty much limited to garages with expensive rates.

Public transit in the city is good, and the Bay Area Rapid Transit system offers good access south on the peninsula and to the East Bay. To enjoy most of the outings listed you should rent a car. Rental agencies are plentiful, and driving is easy once you get out of the immediate downtown area. Taxis are plentiful with reasonable rates.

Indoor Activities

There are visitor centers, museums, interpretive centers, and other indoor facilities available at many of the parks and refuges listed in the Outdoor Activities section that are not listed in this section. Information about these are included in the entries on the various points of interest.

California Academy of Sciences
Golden Gate Park, San Francisco, CA 94118—415-750-7145

This museum has exhibits on the natural history of California, an aquarium, and a planetarium. It is open daily from 10:00 A.M. to 5:00 P.M.

California Marine Mammal Center
Fort Cronkite, Sausalito, CA—415-454-6961

The center shelters and treats injured marine mammals for release into the wild. Visitors are welcome and can observe various exhibits of local marine life. Visiting hours vary.

Coyote Point Museum
Coyote Point Dr., San Mateo, CA 94401—415-342-7755

This museum includes four levels of exhibits on local ecology and environment. Visitors can also hike through a recovering marsh. The museum is open Wednesday–Friday, 10:00 A.M.–5:00 P.M. and Saturday–Sunday, 1:00 P.M.–5:00 P.M.

Junior Museum
199 Museum Way, San Francisco, CA 94114—415-863-1399

This museum has many hands-on nature exhibits that adults can enjoy as well as youngsters. It is open Tuesday–Sunday from 10:00 A.M. to 5:00 P.M.

Lick Observatory
P.O. Box 85, Mount Hamilton, CA 95140—408-429-2495

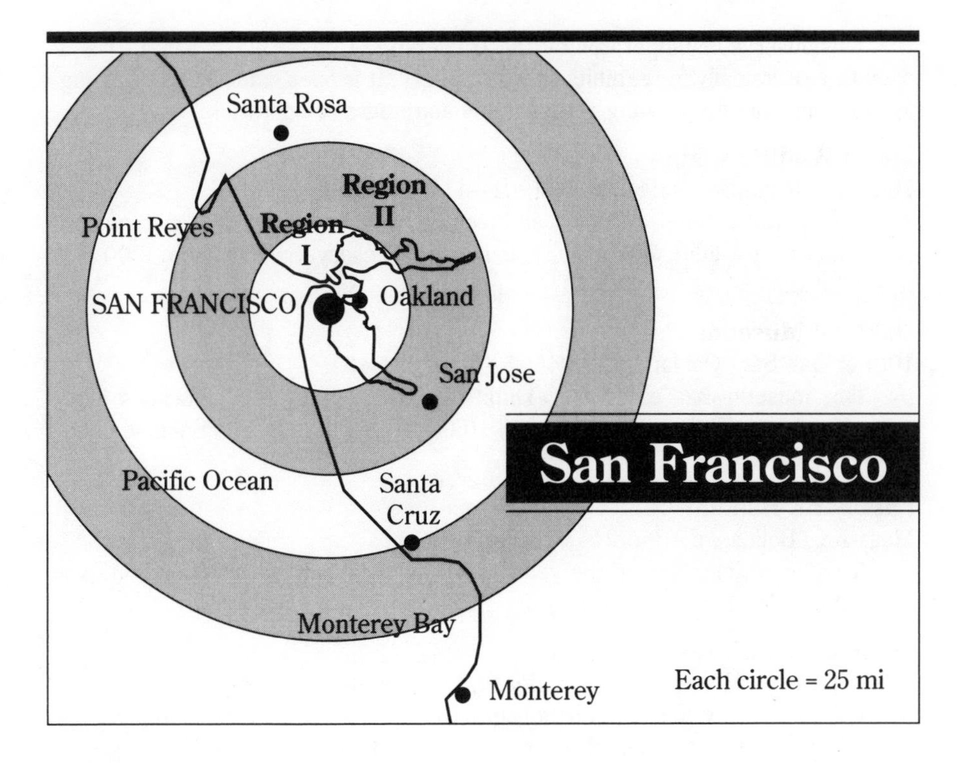

Indoor Activities

Region I

California Academy of Sciences
California Marine Mammal Center
Coyote Point Museum
Junior Museum
Marin Wildlife Center
Oakland Museum
Palomarin Banding Station
Point Reyes Bird Observatory
San Francisco Bay Model
Strybing Arboretum

Region II

Lick Observatory

Outdoor Activities

Region I

Angel Island State Park
Audubon Canyon Ranch
Berkeley Municipal Rose Garden
Charles Lee Tilden Regional Park
China Camp State Park
George Whittel Education Center
Golden Gate Park
Hawk Hill
Marin Water District Lakes
Mount Tamalpais State Park
Purisma Creek Redwoods
Regional Park Botanic Garden
Ring Mountain Preserve
Rotary Natural Science Center
San Bruno Mountain State and County Park
Steep Ravine Environmental Cabins
University of California Botanical Gardens

Region II

Jasper Ridge Biological Preserve
Midpeninsula Regional Open Space District
Point Reyes National Seashore
San Francisco Bay Wildlife Refuge

This observatory, operated by the University of California at Santa Cruz, is open to visitors daily for exhibits and the gallery. It is open some Friday evenings for viewing. Call for viewing reservations and hours of operation.

Marin Wildlife Center
76 Albert Park, San Rafael, CA 94901—415-454-6961

The center includes a shelter and treatment center for injured wild animals and exhibits on wildlife and ecology. It is open Tuesday–Sunday, from 9:00 A.M. to 4:30 P.M.

Oakland Museum
10th & Oak Sts., Oakland, CA 94607—415-273-3401

This museum has extensive exhibits on the ecology of California's biotic zones. It is open Wednesday–Saturday, 10:00 A.M.–5:00 P.M. and Sunday, noon–7:00 P.M.

Palomarin Banding Station
Mesa Rd., Bolinas, CA 94924—415-868-0655

At various times during the year visitors can participate in the activities at the station. Call for more information and visiting hours.

Point Reyes Bird Observatory
4990 Shoreline Highway, Stinson Beach, CA 94970—415-858-1221

This is the only full-time bird observatory in the continental United States. It conducts extensive research on migratory birds of the Pacific. It is open to visitors on an irregular basis. Be sure to call ahead.

San Francisco Bay Model
2100 Bridgeway Blvd., Sausalito, CA 94965—415-332-3870

This is a working model of the San Francisco Bay and its environs. It is used by the U.S. Army Corps of Engineers for research on water movement in and around the bay. Call for experiment times. The model is open Tuesday–Saturday from 9:00 A.M. to 4:00 P.M.

Strybing Arboretum
9th Ave. & Lincoln Way, Golden Gate Park, San Francisco, CA 94118—415-661-1315

More than 5,000 plant species are in the collection here. Outside are 70 acres with demonstration gardens, a Japanese garden, and a Garden of Fragrance for the visually impaired. The arboretum is open daily from 10:00 A.M. to 5:00 P.M.

Outdoor Activities

San Francisco is unique among the cities in this guide in the number of parks, refuges, and reserves that can be considered natural areas, and that are within an hour of downtown. With almost 40 state parks, hundreds of regional and county parks, two national parks, and dozens of natural attractions such as peaks and marshes, it would take a whole book to include information about all of them. To include information about as many as possible, and to give a wide diversity of activities, several types of activities have been combined.

Angel Island State Park

c/o California State Parks, P.O. Box 2390, Sacramento, CA 95811—916-465-6477

This 740-acre island, which sits in the middle of San Francisco Bay off Tiburon, is served by several ferry routes from Tiburon, Vallejo, and San Francisco. The ride from Tiburon is only ten minutes. Once on the island, visitors can spend hours hiking across it to observe both natural and human-made sites.

For years the island was home to military installations, including Nike missile sites in the 1950s and 1960s. Many of the remaining structures can now be toured.

Even with the years of habitation, however, the rugged, brush-covered slopes remained relatively untouched. Today, birds and small mammals native to the region can be found in abundance. Deer, which have long been protected from hunting and which have no natural enemies on the island, frequently experience population explosions. Park officials must remove many of them, rather than have them starve from lack of food.

Trails and unpaved roads lead to all parts of the island, including the top of Mount Livermore, the highest point on the island. Some of the trails are flat and easy, while others area rather steep challenging.

Call for information on the ferry service. It is daily between May and October and periodic during the rest of the year depending on demand and weather. Call 415-435-2131 for service from Tiburon or 415-546-2815 for service from Vallejo and San Francisco.

Audubon Canyon Ranch

4900 Shoreline Highway, Stinson Beach, CA 94970—415-383-1644

This is a popular bird preserve located on the banks of the Bolinas Lagoon. Thousands of people visit the preserve in the spring to see the nesting herons and egrets from viewpoints above the 250-foot firs and redwoods where the nests are located. Many visitors also come to the preserve during the fall migration where thousands of shorebirds invade the lagoon. This outing should not be missed.

Bay Area Ridge Trail

Many hope this ambitious project will complete a 400-mile trail that completely circles the Bay Area. About 200 miles of trails already exist as part of parks and public lands. These are being brought up to standard, and new trails are being constructed to connect those already in place.

The Greenbelt Alliance is the moving force behind this project, and it has maps of the existing trails, as well as those being planned. For more information contact the Greenbelt Alliance, 116 New Montgomery, Suite 640, San Francisco, CA 94105; 415-543-4291.

Berkeley Municipal Rose Garden
Euclid Ave. at Bayview Pl., Berkeley, CA 94708—415-644-6530

A terraced amphitheater and arbor overlook the bay and the Golden Gate Bridge in this garden with over 4,000 varieties of roses. The blooms are especially lush in late spring and early summer. The garden is open during daylight hours.

Brown Pelicans

In 1969 there were only about 1,000 nesting pairs of brown pelicans in California, but by 1985 that figure had risen to more than 6,000. Bird lovers can now view these large sea birds from July through November at a number of spots around San Francisco. Pelicans follow schools of fish but generally can be seen: near the Golden Gate Bridge; at Richardson Bay and Point Bonito in Marin County; on Suisun Bay and the wetlands of San Pablo Bay off State Route 37 near Vallejo, (the northern portion of the San Francisco Bay National Wildlife Refuge); and on the eastern shore of the bay near Hayward (the southern section of the San Francisco Bay National Wildlife Refuge).

Charles Lee Tilden Regional Park
Berkeley, CA 94708—415-841-8732

Part of the extensive East Bay Regional Park system, this 2,065-acre park is set in the hills overlooking Berkeley and the bay. It features hiking trails, a botanic garden, nature area, and an environmental education center. The park is open daily from 8:00 A.M. to dusk.

China Camp State Park
c/o California State Parks, P.O. Box 2390, Sacramento, CA 95811—415-456-0766

This 1,500-acre site was set aside to honor the early Chinese shrimp fishermen who lived here. Over 12 miles of hiking trails are in this small park. They offer access to meadows that are carpeted with wildflowers in the spring, marshes that are home to many shorebirds, and views of the San Francisco Bay. The park is near San Pedro Point in Marin County.

Coast Redwoods

While virgin stands of redwoods once extended both north and south of San Francisco along the coast, there are now only a few virgin stands left for visitors to explore. They are unlike any other place in America. One is Big Basin State Park 20 miles north of Santa Cruz on State Route 1. Call 408-338-6132 for information.

Three state parks and one national monument also feature virgin stands of coast redwoods. Butano State Park is seven miles south of Pescadero; call 415-879-0173. Portola State Park is seven miles west of State Route 35 on State Park Road; call 415-948-9098. Armstrong Woods State Park is on Armstrong Woods Road in Guerneville, California 95446; phone 707-865-2391. Contact Muir Woods National Monument, c/o Golden Gate National Recreation Area, Fort

Mason, Building 201, San Francisco, CA 94123; 415-388-2595. The monument is at the foot of Mount Tamalpais.

East Bay Regional Park District
11500 Skyline Blvd., Oakland, CA 94619—415-531-9300

This park district, which was started in 1934 during the middle of the Great Depression, operates almost 50 units, which are visited by over 15 million people each year. These units are spread throughout the east bay, and all offer hiking, bird watching, picnicking, and nature study.

With so many units space prohibits a description of each, but the EBRPD has a brochure that provides up-to-date information on all units, and notes their special features. To receive this brochure contact the EBRPD. The district also distributes individual folders of the major park units, which include trail maps. You can ask for these if you know which units you plan to visit. During the summer almost all of the units are served by the regional transit system, and the major ones are served year-round.

George Whittel Education Center
376 Greenwood Beach Rd., Tiburon, CA 94920—415-388-2525

The 11-acre Reed Rogers da Fonta Wildlife Sanctuary is part of this Audubon Society center, which offers excellent bird watching, a bookstore with displays of local wildlife, and a lovely Victorian home that has been restored and is open to visitors on Sundays. Interpretive walks are given each weekend.

The 900-acre Richardson Bay Wildlife Sanctuary is also owned by the Audubon Society and is adjacent to the center. A number of trails lead through this sanctuary, and bird watching is excellent.

Golden Gate Park

Golden Gate Park is bordered by the Great Highway, Lincoln Way, Stanyan Street. and Fulton Street. Many activities are available in this large, 1,107-acre, urban park in San Francisco. Bird watching is excellent in and around the many landscaped forests and meadows, and there are miles of foot and bridle paths.

While the park offers a haven in the midst of the city, few visitors realize that the lush vegetation is all the result of John McLaren's pioneering landscape efforts. McLaren was park superintendent from 1887 to 1943. Before he began his tenure, the land now covered by meadows and forests was a barren region constantly being overrun by shifting sand dunes. Through McLaren's efforts the dunes were stabilized, and many exotic and native plants gained a foothold in the sands. Today, such a project would be condemned as ecologically unsound, but this does not lessen millions of visitors' enjoyment.

Hawk Hill
c/o Golden Gate National Recreation Area, Fort Mason, Bldg. 201, San Francisco, CA 94123—415-331-1540

During the fall migration season from October through November, more than 10,000 raptors of 19 species are observed from atop Hill 129 off Conzelman

Road. This is a former military vantage point that has been renamed. Hawk Hill, which sits on a point along the Marin Headlands near the narrowest crossing over the San Francisco Bay, is beneath the concentration point of all migrating hawks that head south down the coastal flyway.

As vast numbers of eagles, hawks, harriers, kites, vultures, ospreys, and falcons pass over the headlands each fall, they are joined by thousands of bird-watchers. They flock to the area to observe what experts say is a concentration of hawks that rivals the most famous of all inland hawk migration routes—the one that passes along Kittatiny Ridge in Pennsylvania where Hawk Mountain is located.

Jasper Ridge Biological Preserve

Stanford University, Palo Alto, CA 94305—415-327-2277

This 1,200-acre natural area is a research preserve owned and operated by Stanford University. Visitors who are interested in flora of the region can join one of the regularly scheduled tours.

Marin Water District Lakes

220 Nellen Ave., Corte Madera, CA 94925—415-924-4600

Marin County depends upon a series of small reservoirs for its water supply, and these eight lakes are excellent hiking sites. Five of the lakes, Phoenix, Lagunitas, Alpine, bon Tempe, and Kent, are located just west of San Rafael at the base of Mount Tamalpais. The other three, Stafford, Nicasio, and Soulejule, are located farther north. All are noted for fishing and for their hiking trails.

Marshlands

Marshes are the nursery and food producer of the estuary ecosystem. Those around the San Francisco Bay have more than 30 plant species that trap nutrients, sift out pollutants, and reoxygenate the water. More than 1.2 million birds avail themselves of the food in the bay each winter on their migration routes, and November is their busiest time. It is also the best time for visitors who wish to see this huge migration effort.

In the past century large portions of the bay marshlands have been filled for building, but a concerted effort has been launched in the past 25 years to save the bay from further degradation. One means of enlisting public support for any such conservation effort is education. More than a dozen sites are now open around the bay where visitors can visit marshlands and learn more about their flora and fauna. These include the following sites.

Martinez Regional Shoreline is on the Carquinez Strait at the end of Ferry Street in Martinez in the northern section of the bay. It has three miles of trails through a 100-acre marsh. Hayward Regional Shoreline at the end of West Winton Avenue in Hayward in the central section of the east bay. It has two miles of trails in a 500-acre marsh.

Other sites include: San Pablo Bay Regional Shoreline, Point Pinole Regional Shoreline, Point Isabel Regional Shoreline, Robert Crown Memorial State Beach,

Shoreline View Park, Miller-Knox Regional Shoreline, Oyster Bay Regional Shoreline, San Leandro Bay Regional Shoreline, Hayward Shoreline Marsh, Palo Alto Baylands, Burlingame Shoreline Park, Coyote Point, Millbrae Sanctuary, Bayfront Park, Paradise Beach, Corte Madera State Ecological Reserve, McNear's County Park, China Camp State Park, and John F. McInnis County Park.

In addition to these parks and preserves there are a number of interpretive centers that help educate visitors about the bay and its ecosystem. These include the San Francisco Bay Model used by the U.S. Army Corps of Engineers and the Coyote Point Museum. See the Indoor Activities section for more information on these.

The Richardson Bay Audubon Center and Sanctuary, off Greenwood Beach Road in Tiburon, features a self-guided walking tour and displays of shorebirds and local wildlife. The Lucy Evans Baylands Nature Interpretive Center on the eastern end of Embarcadero Road in Palo Alto Baylands has seven miles of trails in a 120-acre marsh and an interpretive center.

Crab Cove Visitor Center on McKay Avenue west of Central Avenue in Alameda was the first marine reserve on the bay. It features exhibits on sharks and how the estuary works. Hayward Shoreline Interpretive Center at the end of Clawiter Road off State Route 92 is a 200-acre restored marsh with eight miles of trails. Finally, the San Francisco Bay National Wildlife Refuge Visitor Center on Marshlands Road in Newark has exhibits on endangered species, the wetlands, and winter birding activities. It is at the southern end of the bay, with over 15 miles of trails and a visitor center in 23,000 acres of marshland.

For detailed maps to these marshland sites contact the following agencies: San Francisco Bay National Wildlife Refuge, 1 Marshland Road, Fremont, CA 94536; 415-792-0222; East Bay Regional Park District, 11500 Skyline Boulevard, Oakland, CA 94619; 415-531-9300; Marin County Department of Parks and Recreation, Civic Center Drive, San Rafael, CA 94901; 415-499-6387, and San Mateo County Parks and Recreation Division, 590 Hamilton Street, Redwood City, CA 94063; 415-363-4020.

Midpeninsula Regional Open Space District

Old Mill Office Center, Bldg. C, Suite 135, 201 San Antonio Circle, Mountain View, CA 94040—415-949-5500

One of three such districts in the state, the MROSD has over 22,000 acres that contain some of the best outdoor and natural areas on the peninsula south of San Francisco. These include virgin stands of redwoods, nature study areas for bird-watchers, shoreline parks, and wonderful watershed areas.

Mount Tamalpais State Park

Panoramic Highway, Mill Valley, CA 94941—415-388-2607

Triple-peaked Mount Tamalpais dominates this 6,233-acre park off U.S. Route 101. Hiking trails and winding roads, both paved and dirt, offer visitors opportunities to explore the chaparral growth on the slopes of the mountain. Most of the trails lead to the summit. On clear days, visitors have spectacular views of Marin County, San Francisco, the east bay hills and cities, and the San Francisco Bay. Bird watching is good here, and many mammals such as deer, jack rabbits, and small squirrels can be seen from the trails.

Mountain Peaks in San Francisco Bay Area

Clear winter skies and cool weather give visitors to San Francisco opportunities to gain panoramic views of the region from atop a number of peaks that surround the bay. All of these have well-marked trails to the top that offer hikers a chance to explore the native flora and fauna. Among these peaks are the following.

Mount Diablo (3,849 feet), in Mount Diablo State Park, is 20 miles east of Oakland. On a clear day, this peak offers views from Half Dome in Yosemite to the Farallon Islands. Mount Diablo can be reached by a challenging three-and-one-half-mile trail from the summit parking lot.

Mount Tamalpais (2,571 feet) is in Marin County across the Golden Gate Bridge from San Francisco. It is part of the Mount Tamalpais State Park, which has more than 250 miles of trails.

Mount Livermore (781 feet) is the highest point on Angel Island. It is part of Angel Island State Park and is accessible by a two-mile loop trail to the peak.

Hood Mountain (2,730 feet) is part of Hood Mountain Regional Park in Sonoma County, about 50 miles north of the Golden Gate Bridge. This peak is accessible by a six-mile trail that takes about six hours to hike. On clear winter days, it offers views that stretch from the Sierra Nevada to south of the San Francisco Bay.

Bald Mountain (2,729 feet) and Red Mountain (2,530 feet) are both in Sugarloaf Peak State Park, which adjoins Hood Mountain Regional Park. They can be reached by an eight-mile trail that passes through grassy meadows and scrub oak forests.

Mount St. Helena (4,343 feet) is north of Calistoga in the Napa Valley and has a four-mile trail that climbs 2,400 feet to the peak. Mount Shasta, several hundred miles to the north, is often seen on clear winter days.

San Bruno Mountain (1,314 feet) is in San Bruno Mountain State and County Park, and is home to a number of endangered plant and animal species. A treeless, two-and-one-half-mile trail through grassland leads to the top of the peak, where there are panoramic views of the southern portion of the San Francisco Bay and the east bay hills.

Sweeny Ridge (1,280 feet) is in San Bruno at the southern end of the Golden Gate National Recreation Area. It was from this windy and fog-swept ridge that Gaspar de Portola became the first European to see the San Francisco Bay.

Monte Bello Open Space Preserve (2,000 feet) has a vista point that is reached by an easy three-mile trail. The preserve offers guided hikes along the trail on the third Saturday of every month at 10 A.M.

Vollmer Peak (1,913 feet) and Wildcat Peak (1,250 feet) are both in Tilden Park in the hills above Berkeley. They offer outstanding vistas of San Francisco and the ocean.

Mission Peak (2,517 feet) is in the Mission Peak Regional Preserve in Fremont. It is accessible by a challenging six-mile trail that crosses a number of cattle ranches.

Fall and early winter are the best times to hike these peaks, because the hot summer days have passed, and the winter rains have not turned them to mud. Spring, after the rains have passed and the trails have dried, is another good time, but the air is generally not as sparkling clear then. If possible, avoid hot summer days which cause some trails to close because of fire danger.

Point Reyes National Seashore
Bear Valley, Pt. Reyes, CA 94956—415-663-1092

This is a bird-watcher's and geologist's paradise. This unit of the National Park System has a congressionally designated wilderness area and long stretches of beautiful beaches. Wildlife includes bobcats, deer, coyotes, mountain lions, and tule elk. There are miles of hiking trails, including a self-guided earthquake trail. Some of the best whale-watching sites along the north coast are featured in this park. Abundant bird life thrive in a wide variety of habitats, from seashore to open grassland to upland fir forest. Over 350 bird species have been sighted within the park's boundaries.

The park itself occupies almost all of Point Reyes Peninsula, which is really an island connected by a land bridge to the countryside on the other side of the San Andreas Fault. Although only an hour from downtown San Francisco, this park is one of the finest national parks in the West and an excellent outing for anyone interested in the natural environment.

Purisma Creek Redwoods
c/o Midpeninsula Regional Open Space District, Old Mill Office Center, Bldg. C, Suite 135, 201 San Antonio Circle, Mountain View, CA 94040—415-949-5500

This 2,500-acre preserve could be set in the midst of the great redwood forests hundreds of miles north of San Francisco, but instead, it is only 30 minutes from downtown. It is part of the Midpeninsula Regional Open Space District, and is an undeveloped preserve with no visitor center, rangers, or conveniences, but plenty of ferns, wildlfowers, berries, and beautiful vistas. There are about ten miles of semi-marked trails in the preserve, and the best way to find out about them is to contact the Midpeninsula Regional Open Space District.

Regional Park Botanic Garden
Tilden Regional Park, Berkeley, CA 94708—415-841-8732

This is one of the best exhibits of California native plants in the Bay Area. The garden is open daily from 10:00 A.M. to 5:00 P.M.

Ring Mountain Preserve
c/o Nature Conservancy, 156 Second St., San Francisco, CA 94105—415-777-0487

This preserve is on Esperanza Road in Tiburon and is owned by the Nature Consrvancy. Most residents can direct you to this site, but only when you ask the way to Old St. Hilary's Church. This restored building is the center of activities on the tip of the Tiburon Peninsula, but it is the surrounding hillside that naturalists enjoy. In four acres immediately around the church, more than 220 species of plants grow, including three endangered flowers. These are the Tiburon buckwheat, Tiburon paintbrush, and Marin dwarf flax. The extremely rare black jewel flower is found only in the serpentine soil of the Tiburon Peninsula.

Rotary Natural Science Center

Lakeside Park, Oakland, CA 94612—415-273-3739

During the winter, hundreds of geese, herons, egrets, and ducks take sanctuary at the state wildlife refuge outside the visitor center on the north shore of Lake Merritt. The birds are fed daily at 3:30 P.M.

San Bruno Mountain

State and County Park, P.O. Box 976, Brisbane, CA 94005—415-992-6770

San Bruno Mountain is unique because of its geographical location and specialized weather conditions. More than 20 highly unusual, rare, and endangered forms of plant and animals can be found there. These include 14 plant species, four butterfly species, and one species each of bee, snake, and moth. Early spring, which is during February, is the best time to visit. The higher grasslands burst into vibrant colors with the blossoming of the rare San Francisco Wallflower and Coast Rock Cress.

San Francisco Bay Wildlife Refuge

P.O. Box 524, Newark, CA 94560—415-792-0222

One of the few national wildlife refuges in the midst of a large urban area, the San Francisco Bay National Wildlife Refuge has three units that are open to visitors. The first includes the refuge headquarters, a large visitor center, and over 15 miles of trails that crisscross 25,000 acres of tide flats and marshlands near Newark at the southeast corner of the Bay. The second is the San Pablo Bay unit, which is north of San Francisco, and includes about ten miles of roads and trails but no other facilities. The third is the Farallon Islands unit, which is 27 miles west of the Golden Gate Bridge in the Pacific Ocean. It is accessible only by boat.

Most visitors concentrate on the Newark unit, and miss out on an excellent opportunity to view thousands of waterfowl and shorebirds, plus many raptors, that stop over at the San Pablo unit. Since no visitors are allowed ashore on the Farallons, most visitors enjoy the breeding season by joining one of the tour groups that take boats around the island refuge. Full information on all units is available at the visitor center at Newark.

Steep Ravine Environmental Cabins

Mount Tamalpais State Park, P.O. Box 3015, San Diego, CA 92138-5705—619-452-1950

Just one mile south of Stinson Beach off State Route 1 in Marin County, there are a number of rustic cabins perched on bluffs overlooking the Pacific Ocean. These cabins are open year-round and have no electricity or running water. They are heated by wood stoves. They rent for $25.00 a night and can be booked for up to seven days.

University of California Botanical Gardens
Centennial Drive, Berkeley, CA 94720—415-642-2084

Over 8,000 species of plants, arranged according to their native regions are in this 32-acre garden. There are greenhouse exhibits and a visitor center on the grounds. The gardens are open daily from 9:00 A.M. to 5:00 P.M. Tours are available on Saturday and Sunday at 1:30 P.M.

Wildflower Walks

San Francisco is blessed with an abundance of spring wildflower walks, and with a climate that brings forth these blooms as early as February. Some outstanding walks include the following.

From late February to mid-April on Matt Davis Trail in Mount Tamalpais State Park, calypso orchid, coralroot, fetid adder's tongue, and Indian pipe can be found beneath stands of Douglas fir. Mission bell, white fairy bells, crimson columbine, and Western trillium bloom along the creek beds at the bottom of canyons. Hound's tongue and zygadene open in the mixed woodlands, and poppies and lupine decorate the open meadows.

During March and April on Falls Trail in Mount Diablo State Park, coral-colored wind poppies and short-stemmed purple Chinese houses bloom in open fields beside the lower portion of the trail. Mount Diablo globe lilies and mariposa lilies open along the upper portion, and blue delphiniums bloom in the clefts of the narrow stream as it tumbles down over rocks to form seasonal falls.

From March to mid-April on Summit Loop in San Bruno Mountain State and County Park, yarrow, goldfields, lupine, sticky monkey flower, paintbrush, sun caps, and dozens of other species cover the open hillsides.

During March and April in Edgewood County Park in San Mateo County, Indian warrior, mule ears, lomantium, saxifrage, blow wives, and white fairy lanterns, among others, grow as part of the diverse botanic community on the serpentine soil. For information about Edgewood County Park, contact the San Mateo County Parks, 590 Hamilton Street, Redwood City, CA 94063; 415-363-4020. Information and maps for other wildflower walks can be obtained from the East Bay Regional Park District, and Midpeninsula Regional Open Space District. See previous entries in this section.

Organizations That Lead Outings

The *San Francisco Chronicle* has an Outdoors Section every Monday that covers activities in and around the San Francisco Bay Area. Many organizations are listed in the section, along with the activities they are leading that week.

Bicycle Trails Council of the East Bay
P.O. Box 9583, Berkeley, CA 94709—415-528-2453

Biological Journeys
1007 Leneve Pl., El Cerrito, CA 94530—415-524-7422

California Academy of Sciences
Golden Gate Park, San Francisco, CA 94118—415-221-2100

California Native Plant Society
2380 Ellsworth St., Suite D, Berkeley, CA 94704—415-841-5575

California Native Plant Society
1 Harrison Ave., Sausalito, CA 94965—916-447-2677 (This is the state headquarters in Sacramento.)

Golden Gate Audubon Society
2718 Telegraph Ave., #206, Berkeley, CA 94705—415-843-2222

Marin Audubon Society
P.O. Box 441 Tiburon, CA 94920—415-924-6057

Marin Discoveries
11 First St., Dorte Madera, CA 94925—415-927-0410

Marin Headlands Visitor Center
Golden Gate National Recreation Area, Sausalito, CA 94965—415-331-1540

Nature Conservancy
785 Market St., San Francisco, CA 94105—415-777-0487

Oceanic Society—Bay Chapter
Bldg. E, Fort Mason, San Francisco, CA 94123—415-441-5970

Point Reyes Field Seminars
Bear Valley Rd., Point Reyes, CA 94956—415-663-1200

San Francisco Zoological Society
Sloat Blvd. at Pacific Ocean, San Francisco, CA 94132—415-753-7053

Santa Cruz Mountain Trail Association
P.O. Box 1141, Los Altos, CA 94022—408-968-4509

Sequoia Audubon Society
P.O. Box 1131, Burlingame, CA 94010—415-366-3434

Sierra Club—San Francisco Chapter
6014 College Ave., Oakland, CA 94618 415-658-7470

Whale Center
3929 Piedmont Ave., Oakland, CA 94611—415-654-6621

A Special Outing

With over 21 million visitors a year, the Golden Gate National Recreation Area is the most visited unit administered by the National Park Service. Not all of these visitors come because they are attracted to a great natural area. Many of them are not even aware they have visited the GGNRA, for its attractions are varied and often unmarked. The Golden Gate Bridge attracts millions of tourists each year; the Presidio, a magnet for military enthusiasts, attracts millions more; and Alcatraz Island, the former prison, draws even more. All of these are part of the sprawling Golden Gate National Recreation Area that stretches from Fort Funston south of San Francisco to Point Reyes National Seashore in northern Marin County.

While much of the area is urban, and well developed, there are also natural sites within the recreation area. Most of San Francisco's oceanfront is in the park, and the Marin Headlands are home to an abundance of wildlife and wildflowers. There are miles of uncrowded hiking trails throughout the park. The easy seven and one-half miles of oceanfront trails stretch from the San Francisco Zoo to the Golden Gate Bridge. The more rugged trails crisscross the Marin Headlands. Farther north, Muir Woods National Monument, a unit of the GGNRA, offers visitors opportunities to explore a virgin stand of coast redwoods and enjoy the tranquility it offers.

Whether you are interested in bird watching, wildflowers, sea life, or simply hiking and communing with nature, the Golden Gate National Recreation Area is one of San Francisco's outstanding natural experiences. There is daily transit service to all portions of the unit on the south side of the Golden Gate Bridge, and weekend and holiday service to the Marin Headlands unit. For more information contact the General Superintendent, Golden Gate National Recreation Area, Fort Mason, San Francisco, CA 94123; 415-556-0560, or the Marin Headlands Unit, Golden Gate National Recreation Area, Fort Cronkite, Sausalito, CA 94965; 415-331-1540.

Nature Information

With so many parks and reserves set aside in the San Francisco Bay area there are a number of jurisdictions involved. Only those that have a large number of parks are included in the following list.

San Francisco Visitor Information Center
Powell & Market Sts., San Francisco, CA 94102—415-974-6900

Redwood Empire Association Visitor Center
One Market Plaza, Spear Street Tower, Suite 1001, San Francisco, CA 94105—415-543-8334

Oakland Convention and Visitors Bureau
1000 Broadway Oakland, CA 94607-4020—415-839-9000

California State Parks
P.O. Box 2390, Sacramento, CA 95811—916-445-6477

East Bay Regional Park District
11500 Skyline Blvd., Oakland, CA 94619—415-531-9300

Golden Gate National Recreation Area
Fort Mason, Bldg. 201, San Francisco, CA 94123—415-556-0560

Midpeninsula Regional Open Space District
Old Mill Office Center, Bldg. C, Suite 135, 201 San Antonio Circle, Mountain View, CA 94040—415-949-5500

San Francisco Recreation and Parks Department
Fell & Stanyan Sts., San Francisco, CA 94117—415-558-4431

San Mateo Parks and Recreation Division
590 Hamilton St., Redwood City, CA 94063—415-363-4020

Further Reading

Many of the books listed here have information that pertains to Los Angeles as well as to San Francisco.

Bailey, Edgar H. ed. *Geology of Northern California. Bulletin #190.* San Francisco: California Division of Mines and Geology, 1966.

Bakker, Elna S. *An Island Called California: An Ecological Introduction to Its Natural Communities.* Berkeley: University of California Press, 1972.

Bennett, Ben. *Oceanic Society Field Guide to the Gray Whale.* San Francisco: Legacy Publishing, 1983.

Conradson, Diane R. *Exploring Our Baylands.* 2nd ed. Point Reyes, CA: Coastal Parks Association, 1982.

Gilliam, Harold. *Weather of the San Francisco Bay Region.* Berkeley: University of California Press, 1962.

———. *Island in Time.* San Francisco: Sierra Club Books, 1967.

———. *Between the Devil and the Deep Blue Bay.* San Francisco: Chronicle Books, 1969.

Hart, John. *San Francisco's Wilderness Next Door.* San Rafael, CA: Presidio Press, 1979.

Howard, Arthur D. *Geologic History of Middle California.* Berkeley: University of California Press, 1979.

Howell, John Thomas. *Marin Flora.* Berkeley: University of California Press, 1970.

Liberatore, Karen. *The Complete Guide to the Golden Gate Recreation Area.* Fairfax, CA: Goodchild Jacobsen, 1982.

Margolin, Malcolm. *The East Bay Out: An Unauthorized Guide to Hiking, Swimming, and Fishing in the EBRP's.* Berkeley, CA: Heyday Books, 1974.

McClintock, Elizabeth, Walter Knight, and Neil Fahy. *A Flora of the San Bruno Mountains.* San Francisco: California Academy of Sciences, 1968.

Munz, Phillip A., and David Keck. *A California Flora.* Berkeley: University of California Press, 1973. This is a heavyweight reference work, but Munz also has a series that includes *California's Spring Wildflowers, California's Desert Wildflowers, California's Mountain Wildflowers,* and *Shore Wildflowers.* These books are easier to read.

Newey, Bob. *East Trails: A Guide for Hikers, Runners, Bicyclists, and Equestrians.* 4th ed. Hayward, CA: Footloose Press, 1981.

Reeves, Randall R. *The Sierra Club Handbook of Whales and Dolphins.* San Francisco: Sierra Club Books, 1983.

Rusmore, Jean, and Frances Spangle. *Peninsula Trails.* Berkeley, CA: Wilderness Press, 1982.

———. *South Bay Trails.* Berkeley, CA: Wilderness Press, 1984.

Scanland-Rohrer, Anne, ed. *San Francisco Birdwatching.* Burlingame, CA: Sequoia Audubon Society, 1984.

Sharssmith, Helen K. *Spring Wildflowers of the San Francisco Bay Region.* Berkeley: University of California Press, 1981.

Taber, Tom. *Discovering San Francisco Bay.* San Mateo, CA: Oak Valley Press, 1978.

———. *The Expanded Santa Cruz Mountains Trail Book.* San Mateo, CA: Oak Valley Press, 1982.

———. *Where to See Wildlife in California.* San Mateo, CA: Oak Valley Press, 1983.

Wayburn, Peggy. *Adventuring in the San Francisco Bay Area.* San Francisco: Sierra Club Books, 1987.

Whitnah, Dorothy L. *Guide to the Golden Gate National Recreation Area.* Berkeley, CA: Wilderness Press, 1978.

———. *Guide to Point Reyes National Seashore.* Berkeley, CA: Wilderness Press, 1981.

———. *An Outdoor Guide to the San Francisco Bay Area.* Berkeley, CA: Wilderness Press, 1984.

Both the University of California Press and Wilderness Press have a number of books and series that cover naturalists' activities in California. Most include subjects that affect San Francisco, and many are not included in this list. You can contact the University of California Press at 2120 Berkeley Way, Berkeley, CA 94720; 415-642-4247, and the Wilderness Press at 2440 Bancroft Way, Berkeley, CA 94704; 415-843-8080.

Two periodicals that carry information about nature activities in the San Francisco Bay Area and California in general are the *California Explorer,* which is published bimonthly, and the weekly *Outdoor Section* which is published every Monday in the San Francisco Chronicle.

Los Angeles

Los Angeles is difficult to define. Some have said that it is more a state of mind than of reality, but that is begging the point. Other than New York, there are more people living in the Los Angeles Basin than in any metropolitan region in the United States. To visitors new to the basin it seems that the housing developments and commercial centers are never ending.

It is true that Los Angeles hardly fits the traditional definition of a city. There is no concentrated urban center with distinct limits, and it is often difficult to tell when you have left Los Angeles and entered one of its suburbs. This also makes it difficult to define how far an outing is from the city, for there is no starting point that is likely to be common to most visitors.

ABOVE: These widely spaced valley and coast live oaks and grass-covered hills stand in Cheeseboro Canyon in the Santa Monica National Recreation Area, much of which lies within Los Angeles city limits. Photo courtesy of National Park Service.

Nevertheless, this large urban conglomerate extends from ocean to desert, and has mountains almost a mile high within the limits of its largest city. Los Angeles offers visitors opportunities to explore a wide variety of natural settings.

Southern California is desert country, and lush vegetation is found only at higher elevations or along creeks in shaded canyons. This does not mean there is no diversity in the flora and fauna of the region. The sea is a moderating influence on both temperature and moisture. As you go inland from the seashore, moisture decreases and daytime temperatures rise, until you reach the western boundary of the great Mojave Desert. There rainfall is measured in how many years since the last measurable precipitation fell, rather than in how many inches have fallen in the past year. Summer temperatures frequently go above 110°F.

Comparable changes occur in the flora and fauna as you move inland. Southern California is noted for its fantastic spring wildflower blooms. These stand out more than in other regions of the country because of the lack of competing vegetation. When the blooms come visitors see them in their full glory and color, for there is nothing to detract from their brilliance as they erupt from the brown and green hillsides.

Climate and Weather Information

Many people think Los Angeles has some of the best weather of any major metropolitan region in the country. Rainfall is only about 15 inches a year, and that falls mostly between November and March. Winters are mild with daytime highs averaging between 60 and 75°F. Summers are seldom stifling, with high temperatures generally in the low 80s. Nights are pleasant most of the year.

But Los Angeles does have one great fault—smog. Because of its physical setting, a large basin surrounded by mountains, Los Angeles' climate allows inversions to occur that trap air pollutants near the ground. Also, because there is one car for every 1.8 persons in the region, Los Angeles has the worst air pollution of any metropolitan region in America. This makes breathing difficult at times. Seeing, particularly over long vistas, is impossible, especially during the hot summer months. The clearest days of the year are generally in the fall when hot, dry winds blow in from the desert and sweep the smog out to sea.

Getting Around Los Angeles

This is one city where you probably do not want to consider mass transit. If you want to see any type of attractions you must drive to get there, for that is what the residents do. With the highest ratio of automobiles to people of any region in the world, Los Angeles has become infamous for its traffic and freeways, and rightly so. On some days you can be in stop-and-go traffic for up to 50 miles, and that is not even during rush hour.

The key to getting around is planning. Plan where you want to go, locate the route you want to take on a map, rent a car, and give yourself plenty of time to get there.

The freeway system in Los Angeles is one of the most complex in the world, and visitors are often overwhelmed by the choices and heavy traffic. Los Angeles drivers survive the daily grind of the system. In general, they are more aware of other drivers and more willing to give right-of-way, than those in other major cities. This often gives visitors some chance to correct miscues and missed exits.

Indoor Activities

There are dozens of large and small universities in the Los Angeles area, and many have a variety of natural history exhibits. These are often listed in local papers.

Cabrillo Marine Museum
3720 Stephen White Dr., San Pedro, CA 90731—213-548-7562

This is a showcase for southern California sea life. A 1,100-pound leatherneck turtle, 28-foot skeleton of a young gray whale, and a pile of whalebones you can rummage through are some of the features. A simulated tide pool and over 30 different aquariums are also included.

The museum sponsors special events such as whale watching boat trips and guided tours at the Point Fermin Marine Life Refuge. The museum is open Tuesday–Sunday, from 10:00 A.M. to 5:00 P.M.

Ferndall Nature Museum
5375 Red Oak Dr., Los Angeles, CA 90068—213-467-1661

This museum in Griffith Park has exhibits on the flora and fauna of the park and nearby areas. It offers guided tours, as well as educational programs. Ferndall Nature Museum is open daily from 10:00 A.M. to 5:00 P.M.

George C. Page Museum of La Brea Discoveries
5801 Wilshire Blvd., Los Angeles, CA 90036—213-936-2230

This museum includes exhibits of Ice Age animal remains discovered in the famous La Brea Tar Pits. Visitors can observe workers cleaning and cataloging fossils. The museum is open Tuesday–Sunday, from 10:00 A.M. to 5:00 P.M.

Griffith Observatory and Planetarium
P.O. Box 27787, 2800 E. Observatory Rd., Los Angeles, CA 90027—213-664-1191

On clear evenings visitors can look through a twin refracting telescope. Planetarium shows are featured year-round. Call 213-663-8171 for a 24-hour recorded message on what is going on in the southern California skies. Hours of operation vary seasonally at Griffith Observatory and Planetarium.

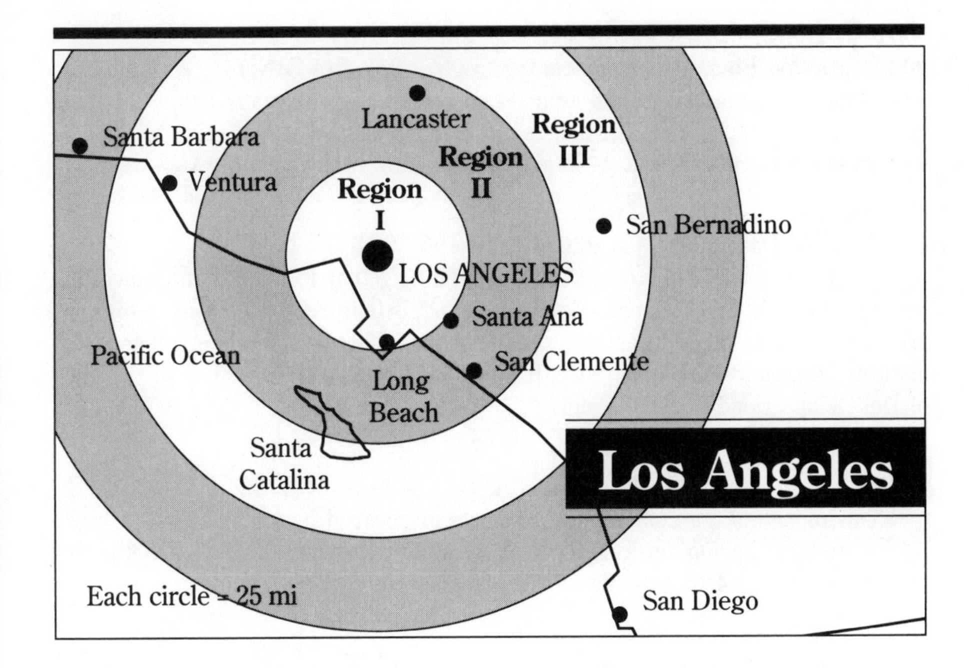

Indoor Activities

Region I
Cabrillo Marine Museum
Ferndall Nature Museum
George C. Page Museum
Griffith Observatory and Planetarium
Natural History Museum of Los Angeles

Region II
Raymond M. Alf Museum

Region III
Orange County Marine Institute

Outdoor Activities

Region I
Descanso Gardens
El Dorado Nature Center
Elysian Park
Griffith Park
Hollywood Reservoir
Huntington Botanical Gardens
Leo Carillo State Beach
Los Angeles County Arboretum
MacArthur Park
Mount Theodore Lukens
Musch Ranch Trail
Point Fermin Park
Point Vincente Interpretive Center
Ralph B. Clark Regional Park
Seal Beach National Wildlife Refuge
South Coast Botanic Garden
Topanga State Park
Upper Solstice Canyon
Whittier Narrows Dam Recreation Area
Will Rogers State Historic Park

Region II
Angeles National Forest
Antelope Coop Trail
Antelope Valley California Poppy Preserve
Antelope Valley Wildflower and Wildlife Sanctuaries
California State Parks (see entries)
Cold Creek Canyon Preserve
La Jolla Canyon
Los Angeles County Parks (see entries)
O'Neill Regional Park
Ocean Vista Trail
Orange County Marine Life Refuges
Placerito Canyon State and County Park
Prado Basin Park
Rancho Santa Ana Botanic Garden
Santiago Oaks Regional Park

Region III
Caspars Wilderness Park

Natural History Museum of Los Angeles County
900 Exposition Blvd., Los Angeles, CA 90007—213-744-DINO

Dinosaurs, gemstones, and animals are all part of the many exhibits found in this huge museum. It is the largest of its kind west of the Mississippi River. The museum is open Tuesday–Sunday, from 10:00 A.M. to 5:00 P.M.

Orange County Marine Institute
P.O. Box 68, Dana Point, CA 92629—714-496-2274

A replica of the *Pilgrim*, the two-masted brig that Richard Dana sailed on and immortalized in *Two Years Before the Mast*, is featured here. Also on display are marine art, nautical treasures, and aquariums filled with live specimens. The institute is open Monday–Saturday, from 10:00 A.M. to 3:30 P.M. It is at the end of Del Obispo Road in Dana Point.

Raymond M. Alf Museum
1175 W. Base Line Rd., Claremont, CA 91711—714-624-2798

This museum is a collection of artifacts discovered by Dr. Alf and his high school students during several paleontological digs. It is open Monday–Saturday from 8:00 A.M. to 4:00 P.M. by advance appointment only.

Outdoor Activities

Los Angeles is so large geographically that the outings listed are spread over much of southern California. Most, however, are within an hour's drive of a major concentration of tourist and business activities. Also, these activities are only a sampling of those available in the Los Angeles area. Visitors who are interested in exploring further will find added attractions along the seashore, in the various national forests around Los Angeles, and in the desert farther afield. There are also many naturalist activities in and around San Diego, which is only two hours or so from the southern portion of Los Angeles County.

Angeles National Forest
701 N. Santa Anita Ave., Arcadia, CA 91006—818-577-0050

With almost 700,000 acres on the northern edge of Los Angeles off Interstate 5, this is the nation's most popular national forest. Located in the San Gabriel Mountains, the forest features spectacular scenery and is home to a variety of wildlife and plants. It was here that the California condor made its last gasp for survival in the wild. Many other seldom-seen animals, including mountain lions, bobcats, foxes, desert bighorn sheep, and black bears can all be found here.

Although this is officially a forest, in many sections visitors see only the low-lying growth of chaparral country. It is only at the higher elevations that fir and pine become dominant.

Three true wilderness areas, and designated as such by Congress, are found in the Angeles National Forest. Only experienced hikers are advised to attempt outings in these rugged regions. Several hundred miles of marked hiking trails are in the forest, including portions of the Pacific Crest National Scenic Trail.

Antelope Valley California Poppy Preserve
c/o California Department of Parks and Recreation, 15101 W. Lancaster Rd., Lancaster, CA 93534—805-945-7811

During April, California poppy, yellow goldfields, purple lupine, and white creamcups blend together to form a multi-hued carpet over the gently rolling hills of this 1,745-acre reserve on the western edge of the Mojave Desert. Many think this is the best time to visit as the desert changes from its subtle brown.

Hiking and nature study are the major activities here, and you may roam freely around the reserve. While most visitors come to see the colorful bloom, the reserve is open year-round, and offers an excellent opportunity to explore the windswept desert region.

An interpretive center is located in the middle of the reserve and features exhibits on the flora and fauna of the region, with special attention paid to the California poppy. The preserve is about 15 miles west of Lancaster off the Antelope Valley Freeway.

Antelope Valley Wildflower and Wildlife Sanctuaries
c/o Natural Area Division, County of Los Angeles Department of Parks and Recreation, 433 S. Vermont, Los Angles, CA 90012—213-738-2961

More than 2,100 acres are included in the eight wildflower and wildlife sanctuaries. From March or April through June these are blanketed with a variety of wildflower blooms whose colors stand out from the green and brown backdrops of the desert hills. The rest of the year visitors may see some of the desert wildlife that is protected by the sanctuaries. These include golden eagles, sidewinders, roadrunners, and desert tortoises. The eight sanctuaries are Butte Valley and Phacelia Wildflower sanctuaries and the Alpine Butte, Gerhardy, Mescal, Big Rock Creek, Tujunga, and Payne wildlife sanctuaries.

California Aqueduct Bikeway
Department of Water Resources, Southern Field Division, P.O. Box 98, Castaic, CA 91310—805-257-3610

This 107-mile bikeway is open to bikers and hikers. It borders the east branch of the California Aqueduct, which brings water from northern California to the Los Angeles region. The aqueduct winds its way through the desert and Antelope Valley with its wildflower displays. It is parallel to sections of the San Andreas Fault, where geological features created by fault activity are evident.

There is no camping along the aqueduct, but there are rest stops at ten-mile intervals, with facilities. There are plans to extend this bikeway nearly 400 miles north to San Francisco.

Caspars Wilderness Park
33401 Ortega Highway, San Juan Capistrano, CA 92675—714-831-2174

This 5,500-acre park has fertile valleys, groves of coastal live oak, and large meadows of spring wildflowers. Development in the park has been kept at a minimum, and it is managed primarily as a natural area. Almost 20 miles of hiking trails are maintained, and visitors can obtain a park guidebook at the park office. It is open daily from 7:00 A.M. to sunset.

Charmlee County Park
c/o Santa Monica Mountains National Recreation Area, 22900 Ventura Blvd., Woodland Hills, CA 91364—818-888-3770

A fine spring display of wildflowers is found in this 460-acre park off the Pacific Coast Highway on Encinal Canyon Road. Sprawling meadows, dense oak stands, and chaparral-covered hillsides are seldom crowded because the park is basically undeveloped. Future plans, including an information center and campground, will make the park more crowded. For now visitors can hike in solitude along the network of trails and fire roads that wind through the park. It is open daily during daylight hours.

Cold Creek Canyon Preserve
c/o The Nature Conservancy, Southern California Chapter Project Office, 213 Stearns Wharf, Santa Barbara, CA 93101—805-962-9111

Located off U.S. Route 101 northwest of Los Angeles in the Santa Monica Mountains National Recreation Area, the last free-flowing creek in Los Angeles County is found in this preserve. Its banks and the surrounding canyons are home to a rich diversity of plant life. There are also interesting geological features.

The 560 acres, owned and operated by the Nature Conservancy, have waterfalls, wind caves, dry rocky ridges, and moist canyon floors. Wildflowers are found in profusion during the season. There is a three-quarter-mile, self-guided nature trail, and another two miles of trails in the preserve. Poison oak and rattle snakes both thrive here. The preserve is open daily year-round, but visitors must apply in advance for a permit or join a guided nature walk.

Descanso Gardens
1418 Descanso Dr., La Canada, CA 91011—818-790-5571

This former private estate has thousands of shrubs and ornamentals that provide blooms year-round, plus a native plant garden. Nature trails lead into wooded areas, and many birds frequent the gardens. Five species of hummingbirds live here, as does the Cooper's Hawk. The gardens are open daily from 9:00 A.M. to 4:30 P.M.

Devil's Punchbowl County Regional Park
28000 Devil's Punchbowl Rd., Pearblossom, CA 93553—805-944-2473

The stunning geological landscape featured in this 1,310-acre park is the result of millions of years of earth movement. Several fault lines, including the San Andreas Fault, have folded, compressed, and broken the rock into bizarre formations. Desert and chaparral wildlife lives in the patches of plants that survive among the jumbles of rocks. A mile-long loop trail and a half-mile nature trail offer hikes for novices, while a longer six-mile hike leads into the adjoining Angeles National Forest. The park is open daily during daylight hours.

El Dorado Nature Center
7550 E. Spring St., Long Beach, CA 90815—213-421-9431

Nature trails wind through the 80 acres of semiwilderness. The nature center has wildlife exhibits, photographs, and artwork. The center is open Monday–Friday, 10:00 A.M.–4:00 P.M. and Saturday–Sunday, 8:00 A.M.–4:00 P.M.

Elysian Park
c/o Griffith Park Ranger Station, 4730 Crystal Springs Dr., Los Angeles, CA 90027—213-665-5188

These 600 acres adjoin Dodger Stadium in Chavez Ravine, but once on the over ten miles of trails that lead through the park, visitors seldom notice the surrounding civilization. Los Angeles' first arboretum is located here, and it has many exotic plants. The park is open daily from 6:00 A.M. to 10:00 P.M.

Griffith Park
Ranger Headquarters, 4730 Crystal Springs Dr., Los Angeles, CA 90027—213-665-5188

Off Interstate 5 northwest of downtown Los Angeles, this is one of the largest municipal parks in the world, with 4,063 acres. Griffith still contains acres of undeveloped land that is unchanged since Indian villages dotted the valleys and hillsides. While there are many human-made attractions in the park, naturalists are more attracted to the 53-mile trail system that leads into the backcountry. The most popular section of this system is the three-mile trip that leads from the parking lot at Griffith Observatory to the top of 1,625-foot Mt. Hollywood and back. There are also some short nature trails near the Ferndell Nature Museum.

Hollywood Reservoir
c/o Department of Water and Power, City of Los Angeles, 111 North Hope St., Los Angeles, CA 90012—213-481-4211

There are no facilities at the reservoir, but there are hiking trails that circle the lake. Few people visit this secluded spot, and the brush-covered hills around it are home to a variety of wildlife that comes to the reservoir for water. Hollywood Reservoir is just off U.S. Route 101 west of Griffith Park.

Huntington Botanical Gardens
1151 Oxford Rd., San Marino, CA 91108—818-405-2100

One of the finest botanical gardens in the world, this 207-acre site contains more than 100,000 specimens of 10,000 species. Even the parking lot, with its avocado and eucalyptus trees and rare herbs and plants, has been called the most attractive in the world. The gardens are open Tuesday–Sunday from 1:00 P.M. to 4:30 P.M. Guided tours normally run Tuesday–Saturday at 1:00 P.M. Sunday visitors must make advance reservations. You can do that by calling or sending a self-addressed, stamped envelope to the gardens.

Leo Carillo State Beach

35000 Pacific Coast Highway, Malibu, CA 90265—805-488-4111

The beach is off State Route 1 or the Pacific Coast Highway in the western part of Los Angeles County. Over 500,000 people a year come to this park, but visitors can still find seclusion on the inland portion, which contains 1,600 acres and climbs to over 1,500 feet in elevation.

The vast majority of the visitors stay on the 6,600-foot beach, and partake of the sun and sea, but that leaves the rugged canyons and brushy hillsides unused. There visitors can hike miles of open trails that lead to fantastic views of the ocean and offer occasional glimpses of coyotes, bobcats, deer, foxes, and rattlesnakes. Bird watching is good here, and gray whales can be seen from the beach from November to May. Tide pools are exposed at low tide near Sequit Point, a bluff that divides the long beach into two separate areas. The beach is open daily from 8:00 A.M. to dusk.

Los Angeles State and County Arboretum

301 N. Baldwin Ave., Arcadia, CA 91006—818-446-8257

This 127-acre arboretum lies next door to the famed Santa Anita Race Track. Both were once part of an 8,000-acre private estate. Today more than 30,000 plants from around the world are growing in the arboretum in a variety of areas, some of which are famous for their use as television and movie settings. "Fantasy Island" and the *African Queen* were both filmed here. The arboretum is open daily from 9:00 A.M. to 4:30 P.M.

Los Angeles Urban Trail System

c/o Los Angeles County Department of Parks and Recreation, 433 S. Vermont Ave., Los Angeles, CA 90020—213-738-2961

Los Angeles has established one of the most extensive urban trail systems in the nation. There are over 225 miles now in operation, with more in the planning stage. Information and maps can be obtained from the Los Angeles County Department of Parks and Recreation.

MacArthur Park

c/o Los Angeles Department of Recreation and Parks, 200 N. Main St., 13th Floor, Los Angeles, CA 90012—213-485-5515

This 32-acre park, near downtown at Alvarado and 6th Streets, has over 107 species of trees. Eighty of them are designated as rare. The small lake attracts gulls and waterfowl from October through March, and many land birds live in the park. It is open from sunrise to 10:00 P.M.

Malibu Creek State Park

28754 Mulholland Highway, Agoura, CA 91301—818-706-8809 (Monday–Friday), 818-706-1310 (Saturday–Sunday)

This popular park is the setting for many films and television shows. It is especially noted for the "M*A*S*H" TV series. There are about 15 miles of trails and fire roads that hikers can use to get into the back areas of the 4,000-acre park, and away from the more popular picnic areas on the edges. Volcanic cliffs, creeks, waterfalls, a marshy lake, and rugged gorges all are found in the park. Animals, from rattlesnakes to an occasional mountain lion, are seen. The park is open daily from 8:00 A.M. to dusk.

Mount Theodore Lukens

c/o Santa Monica Mountains National Recreation Area, 22900 Ventura Blvd., Woodland Hills, CA 91364—818-888-3770

At 5,074 feet, this is the highest peak within the city limits of Los Angeles off Big Tujunga Canyon Road. While the peak offers terrific views on a clear day, the trail to the top is less crowded than many other trails in the city.

Leave from the Doske Picnic Area at the end of Doske Road, and follow Big Tujunga Creek until you spot the Stone Canyon Trailhead on the opposite side of the creek. Cross the creek and follow the trail to the Mount Lukens Fire Road, which leads to the top of the peak. This trail is a little over three miles each way, but the view is well worth it.

O'Neill Regional Park

30892 Trabuco Canyon Rd., Trabuco Canyon, CA 92678—714-858-9365

The northwest corner of this 669-acre park is designated as a natural area, and coyotes, badgers, bobcats, and foxes can be seen there. The park has shrub-covered hillsides, oak and sycamore woodlands, and grassy meadows that are carpeted with wildflowers in the spring. A large area adjacent to the southern boundary of the park is being acquired and will be managed as a wilderness area. This new park will contain the largest stand of California sycamores in the state. The park is open daily from 7:00 A.M. to 10:00 P.M.

Orange County Marine Life Refuges

c/o Environmental Management Agency of Orange County, Recreation Facilities Operations, 10852 Douglass Rd., Anaheim, CA 92806—714-567-6206

Located between Newport Beach to the north and Dana Point on the south are the Newport Beach, Irvine Coast, Laguna Beach, South Laguna, Niguel, Dana Point, and Doheny Beach Refuges. All are accessible from State Route 1, and offer visitors opportunities to explore natural aquariums with a variety of tidepool life.

Placerito Canyon State and County Park

19152 W. Placerito Canyon Rd., Newhall, CA 91321—805-259-7721

This 350-acre park is really the site where the California Gold Rush began, notwithstanding the claims of Coloma far to the north. In 1842, a cattle herder dug up a clump of wild onions to spice up his lunch, and noticed yellow particles clinging to the roots. That find started gold fever in southern California seven years before the more famous rush in northern California in 1849. The park commemorates this event, but also offers over eight miles of hiking trails that lead to waterfalls, mountain ridges, green valleys, and wooded hillsides. Because the terrain is so diversified, a wide variety of flora and fauna thrive in the park. Checklists are available from park rangers. There are also a nature center and two self-guiding nature trails in the park. It is open daily from 9:00 A.M. to 5:00 P.M.

Point Dume State Beach

c/o California Department of Parks and Recreation, Santa Monica Mountains Area Headquarters, 2860-A Camino Dos Rios, Newbury Park, CA 91320—818-706-1310

Sheer cliffs, rich tide pools, remote beaches, and one of the best sites to watch for gray whales are all attractions of this uncrowded beach. It is not accessible by car, but much can be reached by a short hike. The beach is north of Malibu off State Route 1.

Point Fermin Park

807 Paseo Del Mar, San Pedro, CA 90731—213-548-7756

This small, 37-acre park offers one of the best spots to watch migrating gray whales in the Los Angeles area. There are also two trails that lead to the rocky shoreline below where there are several tide pool areas to explore.

Point Vincente Interpretive Center

807 Paseo Del Mar, San Pedro, CA 90731—213-377-5370

An inside observation area and gray whale exhibits are featured at this site. Gray whales can easily be viewed from here during their southern migration. The center is adjacent to the Point Vincente Lighthouse off Palos Verdes Drive.

Prado Basin Park

14600 River Rd., Corona, CA 91720—714-735-7130

This park along the north bank of the Santa Ana River includes nearly 2,000 acres. Less than 200 acres are developed. The entire park is managed as a biological preserve and is one of the few natural riparian areas remaining in southern California. Bird watching is excellent here, and more than four miles of interpretive trails lead through the park. A trail guide is available at the visitor center.

Ralph B. Clark Regional Park

8800 Rosecrans Ave., Buena Park, CA 92621—714-670-8045

Rich fossil beds were exposed here as the California Division of Highways removed approximately 350 million cubic yards of sand and gravel between 1956 and 1973. These beds proved so significant that Orange County bought 85 acres of the land and created a park to preserve the discovery. The area was a large marsh about 10,000 years ago, and the remains of at least 70 different species of Ice Age animals have been identified here. The park is open daily from 7:00 A.M. to 10:00 P.M.

Rancho Santa Ana Botanic Garden

1500 N. College Ave., Claremont, CA 91711—714-625-8767

This garden contains an excellent collection of native California plants and flowers. From February to June is the most colorful time to visit the garden. Guided tours are given Sundays at 2:00 P.M. from March to May. The gardens are open daily from 8:00 A.M. to 5:00 P.M.

Saddleback Butte State Park
c/o California Department of Parks and Recreation, High Desert Area Headquarters, 4555 W. Ave. G, Lancaster, CA 93546—805-942-0662

A forest of Joshua trees, which are really members of the lily family that grow up to 40 feet tall, is the main feature of this 2,875-acre park. Saddleback Butte, a granite mountain, erupts from the surrounding flat desert terrain. Desert tortoises, sidewinders, chuckwallas, roadrunners, golden eagles, and prolific spring wildflowers are other attractions. A short nature trail near the visitor center, and a four-mile round-trip trail that leads to the top of Saddleback Butte offer a chance to explore the area. The park is always open.

San Andreas Fault

The San Andreas is the best known of California's earthquake faults. It is the most conspicuous rift of its kind in the world. It is visible almost the entire way from the Mexican border to Mendocino County north of San Francisco, where it heads out to sea. There are several segments in and around Los Angeles where the results of fault movement are plainly visible.

One of these is just southwest of Palmdale along a highway cut on State Route 14. There buckling of rock formations by tectonic compression is clearly visible, and a nearby plaque shows the fault's location and describes some of its features. Devil's Punchbowl County Park (see entry in this section) is another location where results of the fault's actions can be observed. All along Interstate 5 north of Los Angeles, the freeway cuts through uplifts and other rock shifts that have resulted from earth movement along the fault.

For visitors who wish to explore the San Andreas Fault, one of the best source books is *Earthquake Country: How, Why, and Where Earthquakes Strike In California.* It was published by Lane Publishing Company in 1964. The book is now out of print, but many libraries carry it.

Santiago Oaks Regional Park
2145 N. Windes Dr., Orange, CA 92669—714-538-4440

This 125-acre wilderness park is one of the prime bird-watching areas in Orange County. Over 130 species have been reported. Mountain lions, deer, coyotes, and bobcats have also been seen here. There are several miles of hiking trails, including a loop nature tail near the nature center. The park is open daily from 7:00 A.M. to sunset.

Seal Beach National Wildlife Refuge
c/o Kern National Wildlife Refuge, P.O. Box 219, Delano, CA 93216—805-725-2767

This is one of the largest remaining coastal wetlands systems left in southern California. It lies within the boundaries of the U.S. Naval Weapons Station on Edinger Avenue off State Route 1.

More than 100 bird species live in the habitats provided by tidal creeks, salt marshes, mud flats, and sloughs. Visitors are not allowed in the refuge itself, but can view the wetlands and its inhabitants from adjoining Sunset Aquatic Park, which is open 24 hours a day year-round.

South Coast Botanic Garden

26300 Crenshaw Blvd., Palos Verdes Peninsula, CA 90274—213-377-0468

This 87-acre garden was built on landfill and now has a human-made lake, paths, plant exhibits, and many birds. It is open daily from 9:00 A.M. to 5:00 P.M.

Tapia County Park

844 N. Las Virgines Rd., Calabasas, CA 91302—818-346-5008

This 95-acre park is one of the most popular units of the Santa Monica Mountains National Recreation Area. It offers a refuge for many birds, particularly along Malibu Creek where cottonwoods and willows shade the banks. The park is crowded on weekends and is open daily during daylight hours.

Topanga State Park

20825 Entrada Rd., Topanga, CA 90290—213-455-2465

This 9,000-acre park is almost entirely within the city limits of Los Angeles. It is on Entrada Road off Topanga Canyon Boulevard west of downtown. It features deep canyons, oak woodlands, chaparral-covered slopes, and rugged terrain that goes from 200 to 2,100 feet in elevation. All of this is easily classified as rugged, untouched wilderness, and over 35 miles of trails wind through the park. A one-mile loop nature trail is located near the park headquarters. Year-round, the park is open daily from 8:00 A.M. to 5:00 P.M. and from 8:00 A.M. to 7:00 P.M. on weekends. It may be closed because of fire danger during late summer and early fall.

Vasquez Rocks County Park

10700 W. Escondido Rd., Saugus, CA 91350—805-268-0840

This 745-acre park is managed as a natural area, and features giant sandstone slabs that were tilted into strange formations by movements of the San Andreas Fault. There is no trail system, but visitors may explore any part of the park. It is open daily from 8:00 A.M. to sundown.

Whale Watching

Over 12,000 gray whales migrate from summer feeding grounds in Arctic waters off Alaska. They swim some 6,000 miles to breed and calve in warm, shallow lagoons off the coast of Baja California. On their southern migration they travel less than a half mile from shore most of the way. From November to February whales can be spotted at various points along the California coast.

For information on the migration and the best sites for observing the whales, contact any of the visitor bureaus listed or the American Cetacean Society, National Headquarters, P.O. Box 2639, San Pedro, CA 90731; 213-548-6279. Both sources also have information on expeditions and boat tours that take observers onto the ocean for close-up views of the whales.

Whittier Narrows Dam Recreation Area
1000 N. Durfee Ave., South El Monte, CA 91733—818-444-1872

The easternmost 277 acres of this 1,092-acre park has been set aside as a nature study area. Sitting along the banks of the San Gabriel River are several small lakes, a nature center, a native plant nursery, and a raptor sanctuary.

This recreation area is also the hub of the 225-mile urban trail system managed by the Los Angeles County Department of Parks and Recreation. The Rio Honda River and San Gabriel River Trails run north–south through the area, and connect with other trails in the system. The Skyline and San Jose Trails run eastward from the park. There are also five miles of nature trails within the natural area. It is open daily during daylight hours.

Wildflower Walks

Spring wildflower blooms in the southern California deserts and mountains are some of the most beautiful in the nation. Several sites in and near Los Angeles are well worth the visit if you are in the region between February and mid-June. These include the following.

La Jolla Canyon
Point Mugu State Park, 9000 W. Pacific Coast Highway, Malibu, CA 90265—805-987-3303

Off State Route 1 east of Point Mugu in Ventura County, the Ray Miller Trail leads visitors past a seasonal waterfall on a twisting trail up a canyon. Giant coreopsis tower two to five feet high along the trail in the park. During February, one of the best displays of shooting stars in California can be found nearby. Rattleweed, wallflowers, larkspurs, crimson pitcher sage, Indian warrior, and blue-eyed grass can also be seen. The La Jolla Valley Loop Trail leads to a nice stand of purple needlegrass, California's state grass, which surrounds a large pond that attracts birds and wildlife year-round.

Ocean Vista Trail
Charmlee Natural Area County Park, c/o Los Angeles County Department of Parks and Recreation, 433 S. Vermont Ave., Los Angeles, CA 90020—213-457-7247

The park is on Encinal Canyon Road off State Route 1 in west Los Angeles County. Most of this area is open, slightly rolling meadows that are thick with wildflower blooms during March and April. Fires in 1985 have made the annual bloom outstanding with arroyo lupine, California peony, Catalina mariposa lily, shooting stars, blue larkspur, crimson pitcher sage, paintbrush, rein orchid, and penstemon predominate. A Fire Ecology Trail explains the importance of fire in southern California's chaparral ecological system.

Upper Solstice Canyon

c/o Mountains Conservancy Foundation, 3800 Solstice Canyon #1, Malibu, CA 90265—213-456-7154

Numerous wildflowers can be found in this canyon since its 1,000 foot drop in elevation causes significant changes in habitat. A fire burned the area in 1982, and wild pansy, white yarrow, scarlet bugler, penstemons, and various brodiaeas abound. Woolly blue curls occur in profusion on the chaparral-covered hillsides. The blooms here last from March to May. The canyon is at the end of Corral Canyon Road near Malibu Creek State Park.

Musch Ranch Trail

Topanga State Park, 20825 Entrada Rd., Topanga, CA 90290—213-455-2465

This trail is a four-mile loop that takes visitors through oak woodland, chaparral, and grassland on its 600-foot climb to a ridgetop. It is only one of a number of trails in this 10,000-acre park, but it is particularly colorful from March through mid-May. Showy penstemon, canyon sunflower, bush poppy, black sage, and Chinese houses are only a few of the flowers found here.

Antelope Loop Trail

Antelope Valley California Poppy Preserve, 15101 West Lancaster Rd., Lancaster, CA 93534—805-948-1322, 805-945-7811

See entry in this section for information on this preserve. From March through May, for up-to-date information on the wildflower bloom in southern California, contact the Theodore Payne Foundation's 24-hour hotline 818-768-3533.

Will Rogers State Historic Park

14253 Sunset Blvd., Pacific Palisades, CA 90272—213-454-8212

This park was the 187-acre ranch of the late Will Rogers. It is now restored and open to visitors. Nature and hiking trails traverse the park. One trail leads up a grade with an 1,800-foot elevation gain to Topanga State Park. The Will Rogers State Historic Park is open daily from 8:00 A.M. to 7:00 P.M.

Organizations That Lead Outings

There are dozens of organizations in the Los Angeles area that lead outings. Just a few are shown. For listings of local Sierra Club and the Audubon Society chapters, plus more commercial tour leaders, check in local telephone directories and newspapers.

American Cetacean Society
National Headquarters, P.O. Box 2639 San Pedro, CA 90731—213-548-6279

Cabrillo Marine Museum
3720 Stephen White Dr., San Pedro, CA 90731—213-548-7562

California Natural History Tours
P.O. Box 3709, Beverly Hills, CA 90212—213-274-3025

Natural History Museum of Los Angeles County
900 Exposition Blvd., Los Angeles, CA 90007—213-744-DINO

The Nature Conservancy
Southern California Chapter Project Office, 849 S. Broadway, 6th Floor Los Angeles, CA 90014—805-962-9111

Orange County Marine Institute
P.O. Box 68, Dana Point, CA 92629—714-496-2274

Pacific Crest Club
P.O. Box 1907, Santa Ana, CA 92702. No telephone available. Send a self-addressed stamped envelope for information.

Poppy Preserve Interpretive Association
4555 W. Avenue G, Lancaster, CA 93534—805-259-7721

Redondo Sport Fishing Whale Watching Cruises
233 N. Harbor Dr., Redondo Beach, CA 90277—213-372-2111

The Tree People
12601 Mulholland Dr., Beverly Hills, CA 90210—818-769-2663

A Special Outing

A 150,000-acre wilderness area within an hour's drive of over ten million people is something many people can not quite grasp, but the Santa Monica Mountains National Recreation Area is just that. This unit of the National Park Service covers over half of the mountain ecosystem that stretches from Griffith Park in Los Angeles for 50 miles. It climbs to an elevation of over 3,000 feet before descending into the ocean at Point Mugu State Park.

Along the Pacific, the southern boundary of the recreation area encompasses over 50 miles of coastline, much of it still relatively undisturbed. Inland are deep canyons etched by streams that rush to the Pacific, and slopes covered by a type of chaparral community found only in southern California.

Amidst these, visitors can see some of the complex geology of the region, which was created by faulting, folding, and volcanism along a few of the many fault lines that crisscross southern California. The results of these actions have left steep ravines and jutting ridges with exposed bedrock that is easily read, even by novice geologists.

Southern California's Mediterranean climate nutures plant communities that thrive year-round. They put on a spring show of wildflower blooms on green meadows and chaparral blossoms against brown slopes.

Wildlife is abundant in this region of deep ravines and thick chaparral, which conspire to keep people at a distance. Mountain lions and bobcats roam the mountains in search of deer and small mammals, and coyotes howl at the moonlit skies. Hawks and vultures soar above all. Massive numbers of migrating birds that travel the Pacific Flyway rest on the marshes and lagoons along the coast.

Although much of the recreation area is still undeveloped, and difficult to reach even on hiking trails, there are many opportunities for visitors to explore the rugged country. State and county parks are part of the recreation area. Some of these are included in the previous section. The outer regions of the area are the most accessible, popular, and crowded.

For visitors who are more interested in exploring the backcountry, there are many miles of trails and fire roads. A 55-mile trail following the ridge of the Santa Monica Mountains from the Will Rogers State Park on the east to Point Mugu State Park on the west is being established. Some portions of the trail are already open.

Guided hikes throughout the recreation area are offered on a regular basis, and schedules are posted at the visitor information center. Most of the recreation area is open daily year-round, but some sections are closed to public access during the peak of the fire season. Complete information about the recreation area can be obtained from the Santa Monica Mountains National Recreation Area, 22900 Ventura Boulevard, Woodland Hills, CA 91364; 818-888-3770 (visitor information), or 818-888-3440 (park headquarters).

Nature Information

There are literally hundreds of governmental agencies in and around Los Angeles that are involved with parks, nature centers, museums, and nature reserves. Only the major agencies are included here.

California Department of Parks and Recreation
Santa Monica Mountains Area Headquarters, 2860-A Camino Dos Rios, Newbury Park, CA 91320—818-706-1310

California Department of Conservation
Division of Mines and Geology, 107 S. Broadway St., Los Angeles, CA 90012—213-620-3560

California Desert District Office
Bureau of Land Management, 1695 Spruce St., Riverside, CA 92507—714-351-6394

California State Department of Fish and Game
Region 5, 300 Golden Shore, Long Beach, CA 90802—213-590-5132

City of Los Angeles
Department of Recreation and Parks, 200 N. Main St., 13th Floor, Los Angeles, CA 90012—213-485-5515

Forest Supervisor
Angeles National Forest, 701 N. Santa Anita Ave., Arcadia, CA 91006—818-577-0050

Greater Los Angeles Visitors and Convention Bureau
515 S. Figueroa St., Los Angeles, CA 90071—213-689-8822

High Desert Office
California Department of Parks and Recreation, 43066 N. 10th St., Lancaster, CA 93534—805-945-9173

Long Beach Area Convention and Visitors Council
180 East Ocean Blvd., Suite 150, Long Beach, CA 90802—213-436-3645

Natural Areas Division
County of Los Angeles Department of Parks and Recreation, 433 S. Vermont, Los Angeles, CA 90012—213-738-2961

Santa Monica Convention and Visitors Bureau
P.O. Box 5278, Santa Monica, CA 90405—213-393-7593

Further Reading

Many of the books and series listed in the Further Reading section of San Francisco, particularly from the University of California Press, include information about the Los Angeles region.

Belzer, Thomas J. *Roadside Plants of Southern California.* Missoula, MT: Mountain Press, 1984.

Clark, David L. *LA on Foot.* Los Angeles: Camaro Press, 1985.

Dale, Nancy. *Flowering Plants of the Santa Monica Mountains.* Santa Barbara, CA: Capra Press, 1985.

Daniel, Glenda. *Dune Country: A Guide for Hikers and Naturalists.* Athens, OH: Ohio University Press, 1981.

Gagnon, Dennis R. *Hike Los Angeles.* Santa Cruz, CA: Western Tanager Press, 1986.

McAuley, Milt. *Wildflowers of the Santa Monica Mountains.* Canoga Park, CA: Canyon Publishing, 1985.

Pierson, Robert J. *The Beach Towns: A Walker's Guide to LA's Beach Communities.* San Francisco: Chronicle Books, 1979.

Riegert, Ray. *Hidden Los Angeles & Southern California: The Adventurers' Guide.* Berkeley, CA: Ulysses Press, 1988.

Thomas, Bill, and Phyllis Thomas. *Natural Los Angeles.* New York: Harper and Row, 1988.

Honolulu

The Hawaiian Islands were a tropical wilderness when they were settled by early Polynesian explorers. They retained this quality until the early part of this century, when tourism and agriculture combined to wipe out large portions of wilderness. Today little wilderness remains on the island of Oahu, where Honolulu is located, but other islands, notably Hawaii and Kauai, have large wilderness areas that have been barely altered. With unsurpassed diving, tropical rain forests, active volcanoes, and canyons that match the Grand Canyon of the Colorado in scenic beauty, the islands provide visitors with plenty of naturalist opportunities. This is a side of Hawaii that is little publicized, however, and visitors must actively search for these outings.

Nature often has a dark side, and there are several natural hazards that visitors must be aware of in Hawaii. Earthquakes are frequently felt on all the

---

ABOVE: The Hawaii Volcanoes National Park offers visitors an unequaled opportunity to view volcanic activity, such as this hot lava flowing into the Pacific Ocean. Photo courtesy of National Park Service.

islands, but the major danger from these is falling objects. Most have low Richter-scale readings, and visitors should follow normal earthquake preparedness drills when one occurs.

Volcanic activity on the island of Hawaii, although more spectacular, is even less threatening, because it involves shield volcanoes. During an eruption of a shield volcano, lava flows over the rim of the central caldera or from cracks on the slopes of the cone. On the other hand, explosive eruptions, such as the one at Mount St. Helens, send clouds of cinder, ash, and lava balls into the skies and endanger unwary bystanders.

A more deadly and less expected danger on the islands is that from tidal waves, or tsunamis. These follow earth movement, generally offshore, and are particularly dangerous around low-lying areas where even small waves can inundate large areas of land.

Civil defense authorities on the islands sound attention–alert signals when a tidal wave is expected, and areas likely to be affected are cleared. Maps noting danger areas and emergency procedures are found in the front of local telephone directories, and it is a good idea to at least skim through these upon arrival on an island.

All this makes the islands appear to be fraught with danger, but this is not so. It is true there are natural hazards, as the 1989 earthquake in San Francisco and 1946 tsunami in Hawaii prove. These can cause death and destruction, but no region of the country is more likely to experience such dangers than any other.

Climate and Weather Information

Native Hawaiians had no word for weather, for the islands' subtropical climate is equable throughout the year. Temperatures seldom vary by more than five or ten degrees, ranging from highs in the mid-70s to mid-80s and lows from the mid-60s to mid-70s. It is cooler at higher elevations, where it even snows during the mid-winter months.

Rain falls year-round, with slightly more between October and April. The amount of rainfall varies tremendously by location, as is noted on Kauai. Along the leeward side of the island there is an annual rainfall of less than 20 inches. The middle of the island near Waialeale is the wettest spot on earth, with an average annual rainfall of almost 500 inches.

The sun is fierce year-round on the islands, even when the temperatures are moderate. All visitors should wear hats, sunglasses, and a good sun screen to avoid unpleasant sunburns.

Getting Around Hawaii

On Oahu there is good public transit in and around Honolulu. It extends to the rest of the island. Once on the other islands, however, public transit is almost nonexistent. Car rental agencies are plentiful on all islands, but reservations are desirable and often necessary, especially during the peak tourist seasons.

Indoor Activities

While weather is seldom a reason for staying indoors on the islands, visitors sometimes need a break from exploring the outdoors, or want to study some specific subject inside. In addition to the activities listed, there are interpretive centers in state and national parks where visitors can find out more about Hawaii's natural environment.

Bishop Museum
1355 Kalihi St., Honolulu, HI 96819—808-847-3511

The best collection of information on Hawaii, Polynesia, and Pacific Oceania found anywhere is here. While not all of the exhibits relate to naturalist activities, enough do to make a visit worthwhile. The rest of the exhibits are spectacular. The museum also features a planetarium that gives programs on Polynesian skies, and how early explorers navigated the open ocean. The museum is open Monday–Saturday from 9:00 A.M. to 5:00 P.M.

Kokee Natural History Museum
Kokee State Park, P.O. Box 518, Waimea, Kauai, HI 96752—808-335-5871

Learn about the plants and geology of area through exhibits at this park museum. It is open daily from 10:00 A.M. to 5:00 P.M.

Lyman Mission House and Museum
276 Haili St., Hilo, HI 96720—808-935-5021

A modern museum building here houses extensive exhibits on volcanology, geology, and mineralogy. It is open Monday and Wednesday–Saturday from 9:00 A.M. to 4:00 P.M.

Makiki Environmental and Education Center
2131 Makiki Heights Dr., Honolulu, HI 96822—808-942-0990

This museum has displays of birds, fish, insects, and plants that are indigenous to Hawaii. Outside an easy path with signs and markers leads into a rain forest. The center's hours vary, so it is best to call ahead.

Sea Life Park
Waimanalo, Oahu, HI 96795—808-259-7933

Similar to other commercial sea parks on the mainland, Sea Life Park offers visitors excellent opportunities to view many of the birds, fish, and sea mammals that inhabit the ocean around Hawaii. The park is open daily from 9:30 A.M. to 5:00 P.M.

University of Hawaii
Manoa Campus, Honolulu, HI 96826—808-948-8856

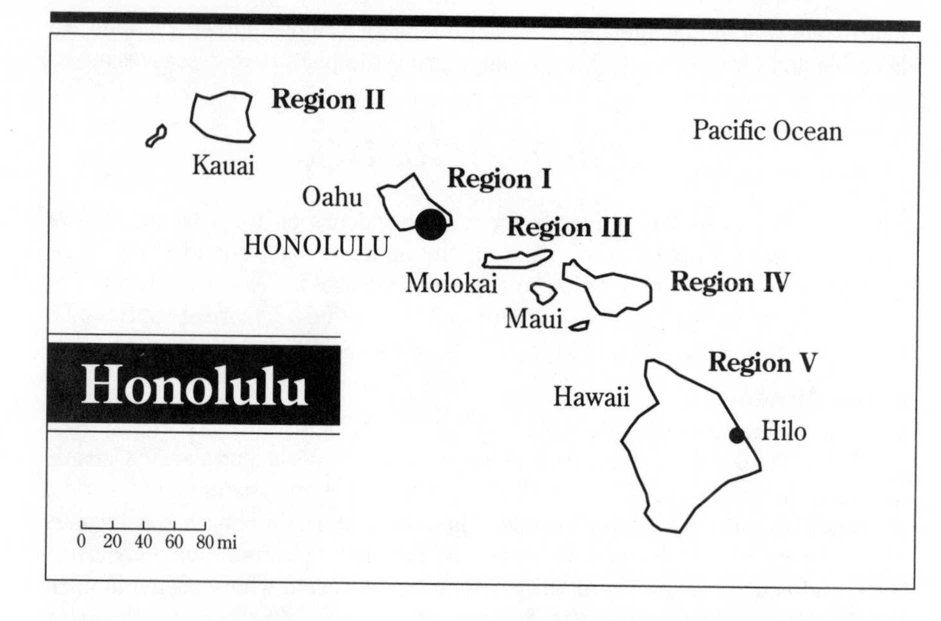

Honolulu

Indoor Activities

Region I
Bishop Museum
Makiki Environmental Center
Sea Life Park
University of Hawaii
Waikiki Aquarium

Region II
Kokee Natural History Museum

Region V
Lyman Mission House and Museum

Outdoor Activities

Region I
Foster Botanic Garden
Hawaii State Parks (see entries)
Honolulu Zoo
Koko Head Park
Liliuokalani Gardens
Lyon Arboretum
Moanalua Valley
Paradise Park
Round Top Forest Reserve
Wahiawa Botanical Gardens
Waimea Falls Park

Region II
Hawaii State Parks (see entries)
Pacific Tropical Botanical Garden

Region III
Garden of the Gods
Halawa Valley
Kamakou Reserve
Moalula and Hipuapua Falls
Molokai Forest Reserve
Palaau State Park

Region IV
Haleakala National Park
Hawaii State Parks (see entries)
Kahana Bird Sanctuary
Kula Botanical Gardens
Maalaea Bay

Region V
Hawaii State Parks (see entries)
Hawaii Tropical Botanical Garden
Hawaii Volcanoes National Park
Kapoho Cone
Kaumana Caves
Kealakekua Bay
Lava Tree State Monument
Nani Mau Gardens

This is the largest campus of the university. It is noted for its departments of oceanography, geophysics, tropical agriculture, and marine biology. Many of these have permanent and special exhibits, which are often noted in local newspapers. Call for hours exhibits are open.

Waikiki Aquarium
2777 Kalakaua Ave., Honolulu, HI 96815—808-923-9741

More than 300 species of Pacific marine life are on exhibit here. The aquarium is open daily from 9:00 A.M. to 5:00 P.M.

Outdoor Activities

From palm-shaded seashore to snow-covered peaks, the islands of Hawaii offer a wide variety of naturalist activities that many visitors never explore. The beauty of these outings is unsurpassed.

Akaka Falls State Park
c/o Division of State Parks, P.O. Box 936, Hilo, HI 96721—808-961-7200

This is a 65-acre park with a paved path that leads through a forest of many different trees and shrubs to two waterfalls, one of which drops 442 feet into Kokekole Stream. Tropical vegetation is so dense along the trail that the sunlight is muted. The park is about 20 miles northwest of Hilo off State Route 220.

Foster Botanic Garden
180 N. Vineyard Blvd., Oahu, HI 96817—808-533-3214

This garden was established by Queen Kalama in 1855, and many of the trees she planted still flourish. More than 4,000 species of tropical plants are found here, and guided tours are offered Monday–Wednesday at 1:30 P.M. The garden is open daily from 9:00 A.M. to 4:00 P.M.

Garden of the Gods
c/o Koele Company, P.O. Box L, Lanai, HI 96763—808-565-6661

Strewn boulders and disfigured lava formations that appear to have been dropped from the sky are found in this sand canyon. It is seven miles outside Lanai City.

Haena State Park
c/o Division of State Parks, P.O. Box 1671, Lihue, Kauai, HI 96766—808-245-4444

The trail head of the difficult 11-mile Kalalau Trail, the only land route into the remote valleys of the Na Pali coast, is in this park. Two large wet caves and the popular Ke'e Beach are also here.

Halawa Valley

This four-mile long and half-mile wide valley is popular with backpackers and hikers. Campers must get permission from the Puu-o-Hoku Ranch and Lodge, Kaluaaha, Molokai, HI 96748; 808-558-8109. It is located at the mouth of the valley.

Haleakala National Park
P.O. Box 369, Makawao, Maui, HI 96768–808-572-9306

More than 30 miles of hiking trails are maintained in this park of cinder cones and wilderness. It is on State Route 378. One of the largest on earth, Haleakala Crater is seven and one-half miles long, two and one-half miles wide, and 3,000 feet deep. Silversword Loop Trail leads through areas covered with the six-foot tall silversword plants. These are outstanding during the summer months when they are in full bloom. The 'Ohe'O Gulch, a rain forest and wilderness area with streams, falls, pools, and bamboo forests, receives over 300 inches of rain a year. The rare nene or Hawaiian goose has been reintroduced to this area, and can be seen around the eastern section of Haleakala Crater.

Harry K. Brown Beach Park
c/o Division of State Parks, P.O. Box 936, Hilo, HI 96721–808-961-7200

Unusual lava formations are found in this park with black sand beaches. It is open daily from dawn to dusk.

Hawaii Tropical Botanical Garden
Hilo, HI 96721–808-964-5233

This is a 17-acre nature preserve that protects a portion of the fragile ecosystem of a tropical rain forest. Trails lead through the preserve, which is also a refuge for shorebirds, forest birds, and giant sea turtles. The garden is open daily from 9:00 A.M. to 4:00 P.M. It is seven miles north of Hilo off State Route 19.

Hawaii Volcanoes National Park
Hawaii National Park, HI 96718–808-967-7311

One of the most active volcano areas in the world is located within the park. Here are the two remaining active volcanoes in the Hawaiian chain. Kilauea is one of the most studied volcanoes in the world, and Mauna Loa is the largest.

The park is open year-round, but some roads and trails are often closed because of continuing volcanic activity. Information about this is available at any of the three visitor centers within the park, or by calling 808-967-7977.

The visitor centers include Jaggar Museum, located three miles west of the Kilauea Visitor Center. The museum is dedicated to seismology and volcanology. The Kilauea Visitor Center on Crater Rim Road is the park headquarters. It contains displays on volcanic land formations. The Waha'ula Visitor Center, near the southeastern entrance to the park on Chain of Craters Road, has a Hawaiian culture museum. There are many points of interest in the park, and information about them can be obtained by calling or writing to the Superintendent.

Honolulu Zoo
151 Kapahulu Ave., Oahu, HI 96815–808-923-7723

One of the world's best collections of tropical birds is found here. The zoo is open daily from 8:30 A.M. to 4:00 P.M.

Iao Valley State Park
c/o Division of State Parks, P.O. Box 1049, Wailuku, Maui, HI 96793—808-244-4352

This cul-de-sac has walls almost a mile high, and is densely forested. The West Maui Forest Reserve includes the valley and surrounding region on the western peninsula of Maui. Iao Valley State Park, Iao Needle, a 1,200-foot rock formation, and Kepaniwai Park are all located here. Iao Valley Road in Wailuku is a continuation of Main Street and follows Iao Stream three miles to Iao Valley State Park.

Kahana Bird Sanctuary
c/o Division of Forestry and Wildlife, P.O. Box 1015, Wailuku, Maui, HI 96793—808-244-4352

About two miles southwest of the airport, this area has an observation hut where migratory ducks, mud hens, stilts, Canada geese, and other birds can be viewed. It is Hawaii's most important bird sanctuary and home to several endangered species.

Kapoho Cone
c/o Division of State Parks, P.O. Box 936, Hilo, HI 96721—808-961-7200

A broad, low-rimmed cone formed by an early volcanic eruption now encloses a six-acre lake. Reflection of the surrounding jungle growth and large amounts of algae make the lake appear green. The cone is on the island of Hawaii, ten miles east of Pahoa on State Route 137.

Kaumana Caves
c/o Division of State Parks, P.O. Box 936, Hilo, HI 96721—808-961-7200

These caves are lava tubes formed by the 1881 eruption of Mauna Loa. This flow came closer to Hilo than any on record. The cavern that is open to the public is about a half-mile long, ten to 50 feet wide, and two to six feet high. The caves are about five miles southwest of Hilo on State Route 200.

Kaumahina State Park
c/o Division of State Parks, P.O. Box 1049, Wailuku, Maui, HI 96793—808-244-4352

Puohokamoa Falls, native plants, and trails that lead to vista points of Keanae Peninsula are all features of this park. Views of farms, the Hana coastline, and the Keanae Peninsula are also provided at this park on the island of Maui in Keanae.

Kealakekua Bay
c/o Division of State Parks, P.O. Box 936, Hilo, HI 96721—808-961-7200

The Captain Cook Memorial was erected in 1874 on this bay near where he was killed attempting to stop a fight between his men and the islanders. The bay itself is designated as a marine reserve because of the clarity of its water and the variety of marine life living there. The bay is on the island of Hawaii in Napoopoo.

Kokee State Park

c/o Lihue District Office, Hawaii Department of Land and Natural Resources, Division of State Parks, P.O. Box 1671, Lihue, Kauai, HI 96766—808-245-4444

Rugged, mountainous terrain is featured in this park in a primitive section of Kauai near Waimea. The Iliau Nature Loop, a serene stroll, and the more strenous Kukui Trail, which leads into Waimea Canyon, are two hiking trails of the park. Many other hiking trails lead visitors into remote areas.

Koko Head Park

c/o Division of State Parks, P.O. Box 621, Honolulu, HI 96809—808-548-7455

This park on Oahu has over 1,200 acres. It includes a botanic garden that emphasizes succulents; volcanic cones; an old lava tube that sometimes causes incoming breakers to spout high into the air; and Hanauma Bay, one of the prime snorkeling spots on the islands. This is a popular and often crowded spot, but the clear waters of the bay and the plentiful fish, makes this an essential outing for any naturalist visiting Hawaii.

Kula Botanical Gardens

Kekaulike Ave., Kula, Maui, HI 96793—808-878-1715

Tropical and semitropical plants are found in these gardens. Paved paths lead by streams, ponds, and waterfalls. The gardens are open daily from 9:00 A.M. to 4:00 P.M.

Laupahoehoe Beach Park

c/o Division of State Parks, P.O. Box 936, Hilo, HI 96721—808-961-7200

This park is on a beautiful peninsula surrounded by spectacular seacoast. The site is most noted for the tidal wave of 1946 that wiped out a village once located here. The park is in Laupahoehoe on the island of Hawaii.

Lava Tree State Monument

c/o Division of State Parks, P.O. Box 936, Hilo, HI 96721—808-961-7200

This monument is in the Nanwale Forest Reserve, and was once the site of a grove of large ohia trees. In the late 1700s an eruption from Kilauea engulfed the grove. Moisture in the tree trunks chilled and hardened the lava into rigid shells of the standing trees. As the molten lava drained away, the hardened shells remained. The monument is three miles east of Pahoa on State Route 132 on the island of Hawaii. It is open daily from dawn to dusk.

Liliuokalani Gardens

c/o Honolulu City Parks, 650 S. King St., Honolulu, HI 96720—808-523-4525

A deep natural stream and a pool are fed by two waterfalls in five acres of unspoiled nature near downtown Honolulu. The gardens are between the Lunalilo Freeway and School Street.

Lyon Arboretum
3860 Manoa Rd., Oahu, HI 96822—808-988-7878

This is a 124-acre rain forest that has many exotic and economically important tropical plants. The environment is usually damp and muddy. The arboretum is open Monday–Friday from 9:00 A.M. to 3:00 P.M.

Maalaea Bay

These are spawning grounds for hump-backed whales, who arrive here around November to give birth. Maalaea Bay is on the island of Maui.

Mauna Kea State Park
c/o Division of State Parks, P.O. Box 936, Hilo, HI 96720—808-961-7200

On State Route 200 about 35 miles west of Hilo, this is a 20-acre park at 6,500 feet. It contains cinder and spatter cone formations and examples of shield volcanism. The road continues on to the summit of Mauna Kea, where snow may be encountered. Skiing is often available between November and February.

Moalula and Hipuapua Falls
c/o Division of State Parks, P.O. Box 153, Kaunakakai, Molokai, HI 96748—808-533-5415

A two-mile trail leads through lush tropical foliage to these two falls. One is 250 feet high and the other is over 500 feet high. The falls are in Halawa on the island of Molokai off State Route 450.

Moanalua Valley
c/o Honolulu City Parks, 650 S. King St., Honolulu, HI 96813—808-523-4525

This valley behind the Moanalua Park and Gardens is managed by the nonprofit group Moanalua Valley Gardens Foundation. There are no developments such as restrooms and water once visitors head into the valley, but this is a favorite nature outing for residents. The valley is off the Lunalilo Freeway.

Molokai Forest Reserve
and
Kamakou Reserve
c/o Molokai District Office, Department of Land and Natural Resources, P.O. Box 153, Kaaunakakai, Molokai, HI 96748—808-533-5415

A 1,750-foot waterfall cascades to the sea over high cliffs on this nearly inaccessible northern watershed region of Molokai. Some of the world's highest sea cliffs, at over 3,300 feet, are found in this coastal region. Few people use the mountain and coastal trails found here. Even fewer live in the isolated countryside. The Molokai Reserve is a state forest and the Kamakou Reserve is owned by the Nature Conservancy.

Nani Mau Gardens
421 Makalika St., Hilo, HI 96720—808-959-3541

One of the largest collections of orchids in Hawaii is found in this state-run arboretum. There are also over 200 varieties of flowering plants, tropical fruit trees, and a Japanese garden. Gardens are open daily 8:00 A.M. to 5:00 P.M.

Na Pali Coast State Park
c/o Division of State Parks, P.O. Box 1671, Lihue, Kauai, HI 96766–808-245-4444

Rugged hiking trails, zodiac boats, and helicopter are all ways to reach the steep cliffs and deep valleys of this remote region in Haena on the island of Kauai. The 11-mile Kalalau Trail is the only access to the region from inland.

Pacific Tropical Botanical Garden
Lawai, Kauai, HI 96766–808-332-7361

This congressionally approved botanical garden offers a collection of endangered tropical plants, particularly those with medicinal and economic importance. The grounds and museum are open Monday–Saturday, from 7:30 A.M. to 4:00 P.M. Guided tours, which last two and one-half hours, are offered at 9:00 A.M. and 1:00 P.M. The garden is off State Route 530 on Hailima Road.

Palaau State Park
c/o Division of State Parks, P.O. Box 153, Kaunakakai, Molokai, HI 96748–808-567-6083

This is a 234-acre forested park on the crest of a mountain that overlooks the Makanalua Peninsula. One trail leads to an overlook that is 1,600 feet above Kalaupapa National Historic Park. It offers outstanding views. A second trail leads to the Phallic Stone, where childless women once came in hopes of becoming fertile. An arboretum with native Hawaiian plants is also located in the park.

Paradise Park
Manoa & East Manoa Rds., Honolulu, HI 96822–808-988-2141

Exotic birds are the main feature of this private park. Beyond it trails lead to a rain forest where, annually, almost 200 inches of rain fall on lush vegetation and feed Manoa Falls and the headwaters of Manoa Stream. The park is open daily from 10:00 A.M. to 5:00 P.M.

Puu Ualakaa State Park
c/o Division of State Parks, P.O. Box 621, Honolulu, HI 96809–808-548-7455

There are many marked trails near the 2,013-foot Tantalus peak that rises above some of the most exclusive homes on Oahu. Of the dozens of outstanding vistas from the mountain, residents claim the best is from the state park.

Round Top Forest Reserve
Oahu District Office, Department of Land and Natural Resources, Division of Forestry and Wildlife, 1151 Punchbowl St., Room 310, Honolulu, HI 96813–808-548-7455

Many trails traverse this forested reserve on Oahu, and several offer spectacular views. Bamboo, Norfolk Island pine, and eucalyptus form a dense canopy for smaller tropical plants.

Sacred Falls State Park
c/o Division of State Parks, P.O. Box 621, Honolulu, HI 96809—808-548-7455

A two-mile trail leads up a rocky mountain ravine to 80-foot falls. The park is off State Route 83 about two miles southwest of Hauula on Oahu. It is open daily from dawn to dusk.

Wahiawa Botanical Gardens
1396 California Ave., Waihiawa, Oahu, HI 96786—808-621-7321

Many varieties of native plants, including rare ferns and trees, are featured here. The gardens are open daily from 9:00 A.M. to 4:00 P.M.

Waianapanapa State Park and Cave
c/o Division of State Parks, P.O. Box 1049, Wailuku, Maui, HI 96748—808-244-4354

The cave is an old lava tube that is reached by swimming underwater from a pool in the park. Many legends surround it. One concerns a Hawaiian princess who hid here from her jealous husband. He later found and killed her. The water near the cave periodically turns red, and the legend claims that is from the blood shed from the murder. Scientists say that it is because of the periodic hatch of millions of tiny red shrimp.

The park includes about two miles of shoreline trails. It is about three miles northwest of Hana on State Route 360.

Wailua River State Park
c/o Division of State Parks, P.O. Box 1671, Lihue, Kauai, HI 96766—808-245-4444

There are six separate recreation areas here, with terrain varying from beach to mountain. There are also several waterfalls accessible by trail. Wailua River State Park is on the island of Kauai.

Waimea Canyon State Park
c/o Division of State Parks, P.O. Box 1671, Lihue, Kauai, HI 96766—808-245-4444

More than ten miles of gorges cut into the Alakai Plateau here. They make a series of brilliantly colored vistas that island residents compare to the Grand Canyon. Several hiking trails take visitors into the gorges. Waimea Canyon State Park is on the island of Kauai.

Waimea Falls Park
59-864 Kamehameha Highway, Oahu, HI 96818—808-638-8511

This is more a tourist park than a nature park, but the 1,800 acres feature an arboretum with over 5,000 species, many of which are endangered. A botanical garden and many birds are also featured. The park is open daily from 10:00 A.M. to 5:30 P.M.

Organizations That Lead Outings

Earthtrust
2500 Pali Highway, Honolulu, HI 96817—808-595-6927

Greenpeace Hawaii
19 Niolopa, Honolulu, HI 96817—808-595-4475

Hawaii Geographic Society
P.O. Box 1698, Honolulu, HI 96806—808-538-3952

Hawaii Trail and Mountain Club
P.O. Box 2238, Honolulu, HI 96814. No telephone available. Send a self-addressed stamped envelope for current hiking schedule.

Hike Maui
101 N. Ihei Rd., Kihei, Maui, HI 96753—808-879-5270

Island Adventure Inc.
559 Kapalulu Ave., Honolulu, HI 96819—808-988-5515

Kauai Mountain Tours
P.O. Box 3069, Lihue, Kauai, HI 96766—808-245-7224

Local Boy Tours
P.O. Box 3324, Lihue, Kauai, HI 96766—808-822-7919

Nature Conservancy
1116 Smith, Suite 201, Honolulu, HI 96817—808-537-4508

Sierra Club Hawaii
1100 Alakea, Room 330, Honolulu, HI 96813—808-538-6616

The Outdoor Circle
200 N. Vineyard Blvd., Suite 506, Honolulu, HI 96813—808-521-0074

Wilderness Hawaii
P.O. Box 61692, Honolulu, HI 96839—808-737-4697

A Special Outing

When volcanoes erupt in most regions of the world they cause people to flee for their safety; when they erupt in Hawaii they attract people. These spectators want a close-up view of one of nature's most awesome events.

The volcanoes of Hawaii often give visitors a chance to view molten lava spewing against the tropical skies. Kilauea, in the Hawaii Volcanoes National Park on the island of Hawaii, is one of the most active volcanoes in the world. Its neighbor, 13,677-foot Mauna Loa, allows millions of cubic yards of lava to escape about every four years.

This is not unexpected, for without this volcanic action the Hawaiian Islands would not exist. The chain stretches almost 2,000 miles across the Pacific Ocean. All of its 124 islands and islets were formed above a hot spot in the earth's mantle over which the Pacific Plate moves. It shifts about two inches a year.

The big island of Hawaii is the latest of the islands to be formed, and is the only one that now has active volcanoes. On several of the other islands, however, there is still plenty of recent geological evidence for visitors to explore.

On Oahu, visitors can find evidence of volcanic activity at the Diamond Head State Monument. A half-mile trail leads from the interior of an extinct crater to its rim. Along the way a variety of rock formations attest to the origin of the island.

Even Hanauma Bay, which teems with sea life and is more known for its diving attractions, is the remnant of a volcanic crater. The view from Koko Head above the bay reveals this.

On Maui, the Haleakala National Park centers around one of the largest volcanic craters on earth. The Haleakala Crater is over seven miles long, two miles wide, and one-half mile deep. An old spatter vent over 60 feet deep is in the center of the crater. The most recent activity, which occurred in 1790, was low on the western slope of the peak rather than in the central crater. Cinder cones up to 700 feet tall are scattered over the crater floor, and visitors can gain a panoramic view of them from the Kalahaku and Leleiwi Overlooks.

Puu Kukui is the older of the two major volcanoes that formed Maui. It has eroded into an almost impenetrable mass of gorges.

But none of the islands offer the amateur volcanologist as much as the big island of Hawaii. There, even the most timid explorer can view molten lava as it bubbles from below the earth's mantle. At times, you can even drive to the very edge of a slow-moving lava flow. More adventurous explorers have labeled Kilauea and Mauna Loa as "drive-in volcanoes" since there is so little danger of their lava welling up from the earth.

Unlike the explosive power released by the eruption of Mount St. Helens, Hawaii's volcanoes erupt with a flow of lava from craters and cracks that offers little danger to residents or tourists. That is not to say that structures and other immovable objects are not in danger. Lava is insistent in its movement toward the sea. Little in its way, either natural or human made, stops it until its temperatures drop, and it returns to solid rock.

This cooling often takes up to a century to complete, however, and visitors can still walk on previous flows and feel the warmth from below. For more information on what to see and do around active and extinct volcanoes contact the individual parks listed in the Outdoor Activities section.

Nature Information

City and County of Honolulu Department of Parks
650 S. King St., Honolulu, HI 96813—808-523-4525

County of Hawaii Department of Parks
25 Aupuni St., Hilo, HI 96720–808-961-8311

County of Kauai Department of Parks
4396 Rice St., Lihue, Kauai, HI 96766–808-245-4982

County of Maui Department of Parks
1580 Kaahumanu Ave., Wailuki, Maui, HI 96793–808-244-5414

County of Molokai Department of Parks
Kaunakakai, Molokai, HI 96784–808-553-3221

Division of Conservation and Resources Enforcement
Department of Land and Natural Resources, 1151 Punchbowl St., Honolulu, HI 96813–808-548-5918

Division of State Parks
Department of Land and Natural Resources, 1151 Punchbowl St., Honolulu, HI 96813–808-548-7455

Hawaii Visitors Bureau
2270 Kalakaua Ave., Honolulu, HI 96815–808-923-1811

Kauai Forest Reserve Trails
P.O. Box 1671, Lihue, Kauai, HI 96766–808-245-4433
Trail lists and maps on Kauai are available.

National Park Service
300 Ala Moana Blvd., Suite 6305, P.O. Box 50165, Honolulu, HI 96813–808-546-7584

Further Reading

Chisholm, Craig. *Hawaiian Hiking Trails.* Lake Oswego, OR: The Fernglen Press, 1989.

Clark, John R.K. *The Beaches of Oahu.* Honolulu: University of Hawaii Press, 1977.

———. *The Beaches of Maui County.* Honolulu: University of Hawaii Press, 1980.

Fielding, Ann. *An Underwater Guide to Hawaii.* Honolulu: University of Hawaii Press, 1987.

Gosline, W.A., and Vernon Brock. *Handbook of Hawaiian Fishes.* Honolulu: University of Hawaii Press, 1970.

Hagman, Marnie. *Hawaii Parklands.* Billings, MT: Falcon Press, 1988.

Hawaii Audubon Society. *Hawaii's Birds.* Honolulu: Hawaii Audubon Society, 1981.

Kuck, Loraine, and Richard C. Tongg. *A Guide to Tropical and Semitropical Floral.* Tokyo, Japan: Charles E. Tuttle, 1958.

Kyselka, Will, and Ray Lanterman. *Maui: How it Came to Be.* Honolulu: University of Hawaii Press, 1980.

MacDonald, Gordon A. *Volcanoes in the Sea.* Honolulu: University of Hawaii Press, 1970.

Manhoff, Milton, and Mitsuo Uyehara. *Rockhounding in Hawaii.* Honolulu: Hawaiian Almanac Publishing Co., 1976.

Merlin, Mark David. *Hawaiian Forest Plants.* Honolulu: Oriental Publishing Company, 1976.

Randall, John E. *The Underwater Guide to Hawaiian Reef Fishes.* Newton Square, PA: Harrowood Books, 1980.

Smith, Robert. *Hawaii's Best Hiking Trails.* 2nd ed. Berkeley, CA: Wilderness Press, 1985.

———. *Hiking Hawaii: The Big Island.* Berkeley, CA: Wilderness Press, 1980.

Stearns, Harold T. *Road Guide to Points of Geologic Interest in the Hawaiian Islands.* Palo Alto, CA: Pacific Books, 1978.

Sutherland, Audrey. *Paddling Hawaii: An Insider's Guide to Exploring the Coves, Jungle Streams, and Wild Coasts of Hawaii.* Seattle: Mountaineers Books, 1988.

Valier, Kathy. *On the Na Pali Coast: A Guide for Hikers and Boaters.* Honolulu: University of Hawaii Press, 1988.

Wallin, Doug. *Diving & Snorkeling Guide to the Hawaiian Islands.* Honolulu: PBC International, 1984.

Index